Houghton Mifflin partners with ETA/Cuisenaire®
to bring you a great value!

A student manipulatives kit specifically designed to accompany

Bassarear's *Mathematics for Elementary School Teachers 2/e*

Each Kit contains:

ManipuLite Pattern Blocks, 27/pkg • ManipuLite Tangram, 7pcs./set • ManipuLite Color Tiles, 20/pkg • ManipuLite Base Ten Flat • ManipuLite Base Ten Rods, 20/pkg • Geoboard • ManipuLite Two Color Counters, 20/pkg • Rubberbands • ManipuLite Base Ten units, 20/pkg • Mesh Bag • Geometric Template (Drawing to Learn)

FOUR EASY WAYS TO ORDER!
For large orders please allow 6-8 weeks for delivery.
- ✉ **Mail:** ETA/Cuisenaire®
  500 Greenview Court
  Vernon Hills, IL 60061-1862
- ☎ **Phone:** 800-445-5985
- ☞ **Fax:** 800-ETA-9326
- 🌐 **Online:** www.etacuisenaire.com

P9-AGU-930

:	Purchase Order #:		ETA/Cuisenaire® Acct #:	Tax Exempt #:

...sure to give street address—UPS Cannot deliver to a P.O. Box)

P TO:	BILL TO: (Type or print clearly)
...ne & Title:	Name & Title:
...ool/Institution:	School/Institution:
...ress:	Address:
:	City:
...e: Zip:	State: Zip:
...norized Signature & Title:	Phone Number:

...antity	Catalog #	Product Description	Price*	Total
	BASSMK	Bassarear Student Manipulatives Kit	$15.95	
	MX	ETA Mathematics Full Line Catalog	FREE	

...ect to change.

...LES TAX: All orders should include state sales tax unless a tax ...ntion number or resale tax number is provided. Sales taxes are required ...states of CA, CT, DC, GA, IL, IN, IA, KS, LA, MD, MI, MN, MO, NC, NJ, ...A, SC, VA, WI, TX and WA. State laws also mandate that sales tax be ...ated on both merchandise and freight.

...PPING CHARGES: All orders from individuals must be accompanied by ...ent or credit card information, and should include 10% for shipping and ...ng charges within the continental U.S.; 20% for shipping charges ...e the continental U.S. Please add $5.00 for shipping and handling on ...under $25.00.

Total Cost of Merchandise	
**Applicable Sales Tax	
†Shipping Charges	
Total (Please allow 6-8 weeks delivery for larger orders)	

❑ **Check Enclosed**   ❑ **Visa**   ❑ **MasterCard**

...# : _____   Expiration Date: _____

...ature: _____

...order form may be reproduced

SECOND EDITION

Mathematics for Elementary School Teachers: Explorations

Tom Bassarear
Keene State College

HOUGHTON MIFFLIN COMPANY   BOSTON   NEW YORK

Sponsoring Editor: *Maureen O'Connor*
Development Editor: *Dawn M. Nuttall*
Editorial Assistant: *Amanda Bafaro*
Project Editor: *Rachel D'Angelo Wimberly*
Editorial Assistant: *Claudine Bellanton*
Senior Production/Design Coordinator: *Carol Merrigan*
Senior Manufacturing Coordinator: *Marie Barnes*
Marketing Manager: *Michael Busnach*

Cover Design: *Minko T. Dimov/MinkoImages*
Cover Image: *Minko T. Dimov/MinkoImages*

PHOTO CREDITS
Chapter 3: p. 86, Plimpton MS 227, f. 14v., Rare Book and Manuscript Library, Columbia University.

Printed in the U.S.A.

Library of Congress Catalog Card Number: 00-133812

ISBN: 0-618-05112-0

3 4 5 6 7 8 9-B-04   03   02   01

CONTENTS

6 ◆ Proportional Reasoning 169

7 ◆ Uncertainty: Data and Chance 183

10 ◆ Geometry as Measurement   325

CUTOUTS

Base 10 Graph Paper
Other Base Graph Paper
Other Base Graph Paper

PREFACE

Having a textbook contain two different volumes created the dilemma of where to put the preface. We resolved this dilemma by having two prefaces. This one, to the Explorations, sets the overall theme of the course and explains how the two volumes work with each other. The preface to the Text more specifically describes the goals of the course and the textbook features. For semantic reasons, this preface is addressed to the student and the Text Preface is addressed to the instructor. However, I have written each with the belief that student and instructor will find both prefaces useful.

   I have been teaching the Mathematics for Elementary School Teachers course for 14 years, and in that time have learned as much from students like you as they have learned from me. The Explorations and the Text reflect the most important things we have taught each other: that building an understanding of mathematics is an active, exploratory process and, ultimately, a rewarding, pleasurable one. Many students have told me and other instructors that this is the most readable and interesting mathematics textbook they have ever read. This is exciting to hear, because I thoroughly enjoy mathematics and hope others will too. At the same time, I know that far from loving mathematics, many people are actually afraid of it. If this book is successful, you will come to believe that math can be enjoyable and interesting (if you don't already feel that way); that mathematics is more than just numbers and formulas and is an important part of the curriculum, and that mathematical thinking can be done by "regular" people.

What is it we want you to learn?

At its heart, the purpose of this course is to revisit the content of pre-K–8 mathematics so that you can build an integrated understanding of mathematical concepts and procedures. From one perspective, it is useful to talk about content knowledge (e.g., different meanings of subtraction, why we need a common denominator when adding fractions, and classification of geometric figures) and process knowledge (e.g., problem-solving and being able to communicate mathematical ideas and solutions). Let me give some examples of the kinds of content knowledge you can expect to learn in this book.

Content Knowledge   Virtually all adults can do the division problem at the right. If asked to describe the procedure, most descriptions would sound something like this: "4 'gazinta' (goes into) 9 two times, put the 2 above the 9, 4 times 2 is 8, 9 − 8 is 1, bring down the 2, 4 gazinta 12 three times with no remainder, put the 3 above the 2, the answer is 23." However, few adults can explain mathematically what "goes into" and "bring down" mean, other than by saying "that's how you do it." However, once you realize that one (of several) meanings of division is that an amount is to be distributed equally into groups and once you realize that 92 can be represented as 9 tens and 2 ones, then you can explain why long division works. Being able to explain *why* gives one what we call mathematical power, which means, among other things, that you can apply this knowledge to solve problems that are not just like the ones in the book. You can solve "real-life" problems, which is one of the most important goals of school mathematics.

   Similarly, most elementary teachers recall that an approximation of π (pi) is 3.14 and they may remember some formulas: $C = \pi d$, $C = 2\pi r$, $A = \pi r^2$, but they can't

$$\begin{array}{r} 23 \\ 4\overline{)92} \\ \underline{8} \\ 12 \\ \underline{12} \end{array}$$

explain what π means or why these formulas work. Developing a
conceptual understanding of π is not as esoteric as many people think.
Let me illustrate. Look at the circle at the right and answer the
following question: Imagine we had several flexible rulers that were
the same length as the diameter. If you wrapped those rulers around
the circle, how many rulers would it take to wrap around the entire
circle? If you actually do this, you find that it will take a little more than three rulers.
Thus, one conception of π is that the length of the circumference of a circle is always a
bit more than 3 times the length of the diameter. With this conceptual understanding,
the formula $C = \pi d$ "makes sense." Ensuring that math makes sense is another key
goal of mathematics education.

 I have just outlined two of many explorations and investigations that you will do
over the course of this book. What so many of my students have discovered is that if
they have a chance to work with mathematical concepts in an active, exploratory
manner, they can make sense of elementary mathematics, which means that they will
be much more effective teachers. Much of the impetus for teaching to foster this style
of learning comes from the National Council of Teachers of Mathematics (NCTM),
which has published three sets of Standards in the past decade and ushered in a new
reform movement in mathematics education. I find that many students who enter the
course with negative feelings toward mathematics view the NCTM Standards very
positively. Many tell me, "I wish this is what my mathematics courses had been like."

NCTM   The first NCTM Standards document, *Curriculum and Evaluation Standards*,
was published in 1989. It sets forth a vision of why mathematics is important for all
citizens to know and describes the mathematical knowledge one should develop by
the end of high school. Chapter 1 of the Text will explore the Curriculum Standards in
more detail. The *Professional Standards for Teaching Mathematics* will be discussed later,
and you should expect to examine the *Assessment Standards for School Mathematics* in
your methods course. Since one of my goals is that you will use NCTM to frame your
own teaching, I will discuss my vision of using NCTM Standards as much as possible.
At appropriate times in the text, I will cite passages from the NCTM documents.

 As I am writing the second edition of these books, the NCTM is finalizing a new
document called *Principles and Standards for School Mathematics* which represents an
update and refinement of the 1989 document. A summary of the ten standards can be
found in Appendix A of the main text, and the full document can be found at the
NCTM website: nctm.org.

What does it mean to learn?

As you will discover in Chapter 1, your attitudes and beliefs have a lot to do with how
you learn mathematics. My beliefs about what it means to "know" mathematics have
led me to create a very different kind of textbook, and it is important for me to describe
my sense of what it means to learn. Let me contrast some traditional beliefs about
mathematical learning (which I disagree with) with this book's approach, and then
describe how this book is structured.

Active versus Passive Understanding   Many students believe that mathematical
understanding is either-or; e.g., you either know fractions or you don't. In actuality,
mathematical understanding is very much like other kinds of knowledge—you "sort
of" know some things, you know other things "pretty well," and you know some
things very well. The belief in either-or leads many students to focus on trying to get
answers instead of trying to make sense of problems and situations. However, when
learning is seen as an "it" that the students are supposed to "get," the student's role is
seen more passively. Our language betrays this bias—the teacher "covers the material"

and the students "absorb." Steven Leinwand, from the Connecticut State Department of Education, once told a joke at a mathematics teachers' convention about the Martian who came to visit American schools. In her report to the Martians, she said, "Teachers are people who work really hard and students are people who watch teachers working really hard!" In a similar vein, the NCTM has a button that reads: "Mathematics is not a spectator sport." Much of the joy of mathematics is examining a situation or problem and trying to understand it. My own experience with elementary school children and my own two children, Emily and Josh, has convinced me that young children naturally seek to make sense of the world that they live in, and that for a variety of reasons many people slowly lose that curiosity over time.

Worthwhile Mathematical Tasks versus Oversimplified Problems   Another common belief about learning mathematics is that we should make the initial problems as simple and straightforward as possible. I call this approach the Lysol approach—we clean up the problems to reduce the complexity (we use routine problems and one-step problems), we try to make the problems unambiguous, and we get rid of extraneous information (we make sure to use every number in the problems). However, one of the biggest drawbacks of this approach is that most of the problems that people solve outside of classrooms are complex and ambiguous, and part of the problem is to determine what information is relevant. A famous Sufi story nicely illustrates the difference between the two approaches. Nasruddin came home one night and found a friend outside on his hands and knees looking in the dirt. When Nasruddin asked, "What's happening?" his friend replied, "I dropped my keys." Nasruddin asked, "Where did you drop them?" His friend pointed: "Over there." Puzzled, Nasruddin asked, "Then why are we looking here?" The response: "Oh, the light is much clearer here." Although the light is much clearer when the problems are unambiguous, routine, and one-step, that is just not how students learn to think mathematically. One consequence of these richer problems (NCTM uses the term "worthwhile mathematical tasks") is that most students report that the problems are much more interesting than what they generally find in texts.

Owning versus Renting   Another way of describing these two very different beliefs about learning mathematics that many of my students have found illuminating is to talk of the difference between owning and renting knowledge. Many students report that one year after finishing a mathematics course, they have forgotten most of what they learned—i.e., if they retook the final exam, they would fail. This means that they rented the knowledge they learned: they kept it long enough for the tests, and then it was gone. However, this need not be so. If you are an active participant in the learning process and if the instructional strategies of the professor fit with how you learn, then you will own most of what you learn. When you take your methods course in a year or so, you will still remember the important ideas from this course. It is such a wonderful feeling to know that you got more than just 3 or 4 credits for the 100+ hours you spent over the course of the semester. If you plan to be an excellent elementary teacher (and I expect you all do), then you need to own what you learn.

An Integrated Approach   If we want students to retain the knowledge, then how they learn it has to be different than it has been in traditional classrooms. One of my favorite words, and one you will find in the NCTM Standards, is *grapple*. I have found that if students grapple with problems and situations and try to make sense of them, they are more likely to retain what they learn in the process. I believe that a central part of the teacher's job is to select worthwhile tasks (Professional Standard 1) and to develop an environment that invites all students to learn and that honors differences in how they learn (Professional Standard 5). A classroom consistent with the NCTM vision does not look like a traditional classroom in which the teacher mostly lectures and demonstrates, and students generally take notes, ask questions, and answer questions the teacher asks. Rather, the classroom looks like an ongoing dialogue: the teacher presents a problem, possibly a brief lecture, and then the teacher facilitates the discussion around that question or situation (Professional Standards 2, 3, and 4), during which students naturally expect to make guesses (predictions and hypotheses) and try to explain their thinking and justify their hypotheses. Similarly, I believe that the textbook for this classroom must be different from a traditional book in which the important concepts and formulas are highlighted and in which the problems at the end of the chapter are generally pretty much like the examples.

In the introduction to NCTM's *Professional Standards for Teaching Mathematics*, the authors summarize Major Shifts in mathematics classrooms that have been called for by the NCTM. This passage nicely summarizes much of what I have just described.

We need to shift:

- toward classrooms as mathematical communities—away from classrooms as simply collections of individuals;

- toward logic and mathematical evidence as verification—away from the teacher as the sole authority for right answers;

- toward mathematical reasoning—away from merely memorizing procedures;

- toward conjecturing, inventing, and problem-solving—away from an emphasis on mechanistic answer finding;

- toward connecting mathematics, its ideas, and its applications—away from treating mathematics as a body of isolated concepts and procedures.

If we can convince our students that fundamentally mathematics is a sense-making enterprise, then we change not only how much they work but also *how* they work. That is, if they see the relevance of the problems and concepts and develop confidence, they work harder; if they see the goal as sense making versus "getting it," they work differently. With this brief overview of what I mean by learning, let me explain how the two volumes work together.

Role of Explorations and Text

As just mentioned, if we believe that people learn better by grappling with richer problems (Worthwhile Mathematical Tasks) as opposed to being shown how to do simpler problems, then this creates a whole host of changes in what the classroom looks like, what the instructional materials look like, and how students' knowledge is assessed. Basically, these Explorations have been designed to have you grapple with important mathematical ideas. I have posed questions and tasks that call on you to discover fundamental mathematical concepts and structures, for example, that percents can be seen as proportions. In the Explorations, you grapple with new ideas and concepts in a hands-on environment. The Text then serves as a resource: a place where the formal definitions and structures are laid out (to be used generally after the Explorations), a place to refine your understanding, a place to self-assess, and of course, a source of homework problems. The Explorations in this volume are generally open-ended and often have more than one valid answer, whereas the Investigations (in the Text) are generally more focused in scope and, though they generally have one correct answer, the discussions following the Investigations show that they always lend themselves to different ways to arrive at that answer. I generally begin each new chapter with an Exploration that, in learning theory terms, enables and encourages students to construct a scaffolding of major mathematical concepts. I then use the Text as a resource where the students can see the concepts articulated in an organized way.

Looking back is a habit of mind that I want to encourage. A famous mathematics teacher proverb is "you're not finished when you have an answer; you're finished when you have an answer that makes sense to you." I model this notion explicitly in Chapter 1 in the Explorations and in the Text and you'll see it often in both volumes. The idea of reflection is new to many students, but getting into the habit of looking back is a great way to take ownership of this content. With this in mind, I hope you're looking forward to exploring elementary mathematics.

Changes from the first edition

- Throughout, 13 Explorations were removed and 16 new Explorations were added. Many other Explorations were revised. Note that all removed Explorations will be available on the website (http://college.hmco.com).

- The Explorations for Chapters 8 and 9 underwent the biggest changes. In both chapters, the first three Explorations are not assigned to specific sections, but instead, cut across topics from the chapter in order to address that chapter's *big ideas*. Each of the three Explorations involve a particular manipulative type: **geoboards, tangrams** or **polyominoes.** These manipulative-specific Explorations make it easier for the instructor to use a single type of manipulative to explore topics from the whole chapter.

- The format of this manual has been completely revised to make it more accessible and user-friendly. First, each Exploration now starts at the top of a new page. Second, the extra answer space, which is never sufficient, has been deleted. Third, wherever appropriate, special pages are provided, containing a table, figure, grid, etc., that would be especially helpful for students to tear-out, fill-in, and either pass in or add to their notes. These special pages are always right-hand pages that are blank on the reverse side.

- The heavy-paper cutout pages at the end of the manual have been updated to include more pages of useful cutouts.

Foundations for
Learning Mathematics

You are about to begin what we hope is a very exciting course for you—a course in which you will come to appreciate the importance and relevance of elementary mathematics in our society, and a course in which you will come to a deeper understanding of the mathematical concepts behind the many mathematical procedures you know. For example, you will discover why we "move over" in the second row when multiplying large numbers, why we find a common denominator when adding fractions but not when multiplying fractions, what π means, and much more.

One of the major differences between this text and other mathematics texts arises from a growing belief that students learn mathematics best by using explorations to investigate and understand mathematical concepts, ideas, and procedures, rather than merely having the teacher explain the concepts, ideas, and procedures. Both volumes have been designed to present you with what the National Council of Teachers of Mathematics (NCTM) has called "worthwhile mathematical tasks."

There are many characteristics of worthwhile mathematical tasks upon which most teachers would be likely to agree. Let us discuss a few. A worthwhile task contains significant mathematics that will require the students to think (as opposed to memorize or follow a procedure). As a result of working on a worthwhile task, students should find that they have come to a deeper, richer understanding of a mathematical concept. Depending on the task, this might mean that they understand a mathematical concept more fully, that they understand why a mathematical procedure works, that they see connections they hadn't seen before, or that they have added to their problem-solving repertoire, to name but a few possibilities. Two other characteristics of worthwhile mathematical tasks concern the students. First, to be worthwhile, the task must be interesting to the students. Second, the task should be rich enough so that all students can be challenged by the task and yet be able to feel some success in the doing of the task. To the extent that a course is rich in worthwhile mathematical tasks, to that extent the students will come to feel more positive toward mathematics, they will acquire confidence in being able to do mathematics, and they will believe that mathematics can and does make sense to "regular" people.

To the extent that these explorations are successful, not only will you learn quite a bit of what is traditionally called mathematical content, but you will also develop more competence with respect to the abilities articulated in the five process standards found in *Principles and Standards for School Mathematics* published by the National Council of Teachers of Mathematics in 2000:

- *Problem solving*   You will become a more powerful and more confident problem-solver.

- *Reasoning and Proof*   Your ability to use reasoning—deductive, inductive, and intuitive—will grow, and you will be able to explain your solution paths.

- *Communication*   You will appreciate the role of discussion in learning mathematics, and you will appreciate the value of vocabulary and notation as tools that make communication easier.

- *Connections*   You will be more aware of connections between various mathematical topics and of connections between mathematics and other areas.

- *Representation*   You will increase your ability to represent problems in effective ways.

As you do these explorations, please keep the NCTM standards in mind, especially the five standards just mentioned, and use the "4 Steps for Solving Problems" guidelines on the inside front cover of this book to help you when you get stuck. When you take your Methods course later in your studies, you will examine how to develop lessons that are consistent with the NCTM standards. To the extent that you experience the standards as a *learner* of mathematics, to that extent you will be able to use the standards more skillfully as a *teacher* of mathematics. At the end of each exploration, you will be asked to "look back"; this activity will help you to reflect on what you have learned.

EXPLORATION 1.1   **How Many Handshakes?**

More and more teachers are having students solve problems in small groups. This method of instruction, called cooperative learning, works better if the students know one another by name.

Determine how many students are in your class, including yourself. The question we will investigate is as follows: If each student shakes hands with every other student in the class, how many handshakes will there be?

1. Write your ideas about how you might solve this problem. Can you apply ideas discussed in the Preface to find patterns in this problem? Can you use what you see to help you plan a solution?
2. Discuss your ideas in your small group or with the whole class, as your teacher directs. Write down new ideas that you like.
3. Now solve the problem. Keep in mind the "4 Steps for Solving Problems" guidelines from the inside cover of this book, and refer to them as needed by asking yourself the following questions: Do you understand the problem? Do you have a plan? Are you monitoring your work? Did you check your solution?
4. After the class discussion of the problem, select a solution path that was different from the one you used. Explain that way of doing the problem, as though you were speaking to a student who was not in class today.

EXPLORATION 1.2   **Savings with Different Coffee Filters**

This is a real problem that confronted some friends of mine. They had just bought a new coffee maker that makes individual cups of coffee. They both liked different flavors of coffee, so now they could each have one cup of their own coffee in the morning. There are specific filters, called funnel filters, for this kind of coffee maker, as opposed to the circular basket coffee filters for the coffee makers that make many cups. At the supermarket, the wife noticed that the basket coffee filters seemed much less expensive and wondered how much money they would save if they used the basket coffee filters. She also noticed that generic filters seemed quite a bit less expensive than name-brand coffee filters.

Compare the yearly cost of using name-brand funnel coffee filters to that of using generic basket coffee filters. You may use the information below, or your instructor may supply you with information from your local supermarket.

Given information:

Brand-name funnel coffee filters—40 for $1.50
Generic funnel coffee filters—40 for 99¢
Brand-name basket coffee filters—100 for $1.29
Generic basket coffee filters—100 for 77¢

1. Work on this problem, alone or in groups, according to your instructor's directions.
2. Share your solutions with your partners or other groups. Your description of your solution should include both the assumptions that you made in order to solve the problem and the actual amount of savings that you determined.
3. Make up and solve a similar problem in which you compare the cost of two or more items.

EXPLORATION 1.3   **Playing Darts**

Problems like this are commonly found in elementary school textbooks. I used this with a group of fifth graders with whom I worked once a week during the 1998–1999 school year.

1. If you have four darts and the dart board shown, which of the scores below are possible scores, assuming that every dart hits the dart board? Be sure to show your reasoning, in addition to your answers.

    6   10   13   15   20   28

2. Can you predict what kinds of scores are possible and what kinds of scores are not possible? For example, do you think a score of 18 is possible or not? Why? (Note the importance of the NCTM Reasoning standard here.)

3. **a.** List all possible scoring outcomes, such as (7, 7, 7, 7), (1, 3, 5, 7), and so on.
    b. Describe your strategy. (Note the importance of the NCTM Communication standard here: You are being asked to *communicate* the *problem-solving* strategy that you used.)

4. What if you had only three darts? What scores would then be possible?

5. Make up and answer your own "What if . . ." question about a dart board. You might use a different number of darts or different numbers for scores such as (1, 5, 10, 25).

EXPLORATION 1.4   **Master Mind**

We will play a variation of a game called Master Mind that has entertained people for many years. I have seen this game in many elementary school classrooms. Teachers have used it to develop logical thinking, and it is a popular choice in classrooms where students are given a chance to spend some time on any activity they choose.

Directions for playing the game:

- One group will think of a four-digit number.
- The second group's goal is to guess the number.
- After each guess, the first group will tell how many digits are correct *and* how many digits are in the correct place. For example, if the number is 1234 and the guess is 2468, the feedback will be two correct digits and none in the correct place.
- Your instructor may limit the game at first, requiring, for example, that players use only the digits 1, 2, 3, 4, and 5 with no repetitions; you can't have 3333, for example.

1.   Play the game several times.
2.   Answer the following questions within the parameters of your game.

    a. If you are in the first group, what number or sets of numbers are you more likely to select, assuming you want to make the number harder to solve? Why?

    b. If you are in the second group, what number would be a good first guess? Why?

    c. Let's say you know some students in another class who haven't played the game yet. What hints would you give them?

EXPLORATION 1.5   **Magic Squares**

Have you ever seen a magic square? Magic squares have fascinated human beings for many thousands of years. The oldest recorded magic square, the Lo Shu magic square shown in the figure below, dates to 2200 B.C. It is supposed to have been marked on the back of a divine tortoise that appeared before Emperor Yu when he was standing on the bank of the Yellow River.

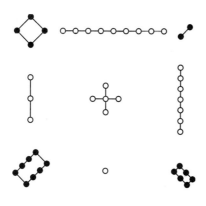

In the Middle Ages, many people considered magic squares to be able to protect them against illness! Even in the twentieth century, people in many countries still use magic squares as amulets.

We will spend some time exploring magic squares because they reveal some amazing patterns. As a teacher, you will find that most students love working with magic squares and other magic figures. Many elementary teachers have their students explore magic squares partly because they are a lot of fun, but also because all five of the NCTM process standards emerge nicely from these explorations.

PART 1: Describing magic squares and finding patterns

1. Let us begin with the simple magic square shown in the figure. Take a couple of minutes to write down why this is a magic square and to write down all the patterns you see in this square.

8	1	6
3	5	7
4	9	2

2. Compare your observations with those of others in your group. Add to your list new observations and patterns that you heard about from other members.
3. Based on your observations and discussion, what makes a magic square "magic"? Imagine telling this to a friend who has never heard of magic squares.
4. Take a few minutes to think of questions you might have about magic squares. Write down these questions.
5. Listen to the questions that other students suggest. Your instructor may select additional questions to investigate.

PART 2: Patterns in all 3 × 3 magic squares

Use the 3 × 3 magic square templates on page 11.

1. Make a completely new 3 × 3 magic square. You are not restricted to consecutive numbers, though I suggest restricting yourself to positive whole numbers, simply to make it easier to see patterns that are true in all the magic squares.
2. Display each magic square. Have each person explain any strategies that made creating the square easier,—that is, strategies beyond "grope and hope." Note any strategies that you did not discover but that you would like to remember.
3. Write down patterns that seem to be true for all 3 × 3 magic squares and patterns that seem to be true only for some 3 × 3 magic squares. (*Hint:* There are lots of patterns!) Then go around the group and have members share their responses.
4. Select one non-simple pattern that is true for all 3 × 3 magic squares.

   a. Describe this pattern, using only words. Imagine that you are talking on the phone to a friend who has seen magic squares but didn't realize there are patterns that are true for all 3 × 3 magic squares.
   b. Give this description to some friends who are not in this course. Check to see how well they understand the pattern just from reading your description. If they understand the pattern, fine. If they don't, revise your description and repeat the process. In the latter case, write down what you learned about communicating.

PART 3: Using algebra to describe 3 × 3 magic squares

1. As you may have already discovered, the middle number of the 3 × 3 magic square is a key number. What if we called the middle number m? Reflect on the patterns you observed in all magic squares and the strategies students used to construct magic squares. Can we represent the other numbers in the magic square in terms of m? With the group working together, take a few minutes to think about this question, and look at relationships between the middle number and other numbers in the sequence. Describe your present thinking and work before moving on.
2. One key to solving this problem is to realize how many more variables are needed. The only complete algebraic representation of 3 × 3 magic squares that I know of requires two more variables. Let us call them x and y. That is, using three variables, we can state instructions for making any 3 × 3 magic square. In other words, by using m, x, and y, we can represent the value of each cell in the magic square. Work on this problem in your group. Briefly explain, in writing, how you solved the problem.
3. At this point, you have a solution, generated either from your group or from the whole-class discussion. Think about your description of how to make a generic magic square, and then read the following quote. Does this experience change your attitude toward algebra? Does it help you to see the use of symbols in a new light? Briefly explain your response.

   Mathematics is often considered a difficult and mysterious science, because of the numerous symbols which it employs. . . .[T]he technical terms of any profession or trade are incomprehensible to those who have never been trained to use them.

But this is not because they are difficult in themselves. On the contrary they have invariably been introduced to make things easy. So in mathematics, granted that we are giving any serious attention to mathematical ideas, the symbolism is invariably an immense simplification.[1]

PART 4: Further explorations for 3 × 3 magic squares

1. How many different 3 × 3 magic squares can you make starting with the two numbers shown in the figure?

3		
		7

2. Below are questions about four possible transformations of a 3 × 3 magic square. In each case, record your initial guess and your reasoning before you test your guess. Then test your guess. If you were correct, refine your justification if needed. If you were wrong, look for a flaw or incompleteness in your reasoning. If you were wrong, can you now justify the correct answer?

   a. If you doubled each number in a magic square, would it still be a magic square?

   b. If you added the same number to each number in a magic square, would it still be a magic square?

   c. If you multiplied each number in the magic square by 3 and then subtracted 2 from that number, would it still be a magic square?

   d. If you squared each number in the magic square, would it still be a magic square?

3. Look back on your data from different 3 × 3 magic squares, and answer the following questions:

   a. Is there a relationship between the magic sum and whether the number in the center is even or odd?

   b. Divide the set of magic squares into two subsets: those in which the nine numbers are consecutive numbers (for example, 10–18) and those in which the nine numbers are not consecutive numbers. Are there any other differences between these two sets of magic squares?

PART 5: 5 × 5 magic squares

1. Three 5 x 5 magic squares are shown below. What do you see (observations and/or patterns)?

17	24	1	8	15
23	5	7	14	16
4	6	13	20	22
10	12	19	21	3
11	18	25	2	9

11	10	4	23	17
18	12	6	5	24
25	19	13	7	1
2	21	20	14	8
9	3	22	16	15

22	29	6	13	20
28	10	12	19	21
9	11	18	25	27
15	17	24	26	8
16	23	30	7	14

[1] Alfred North Whitehead, *Introduction to Mathematics,* (New York, 1911, pp. 59–69), cited in Robert Moritz, *On Mathematics* (New York: Dover Publications, 1914), p. 199.

2. Describe similarities and differences that you see between 3×3 and 5×5 magic squares.

3. Use the 5×5 magic square templates on page 11. Your instructor will walk you through a set of instructions for generating 5×5 magic squares. Your task is to make sense of these rules and then write out a set of rules that you could email to a friend who needs them to make 5×5 magic squares successfully.

3 × 3 Magic Square Templates For EXPLORATION 1.5

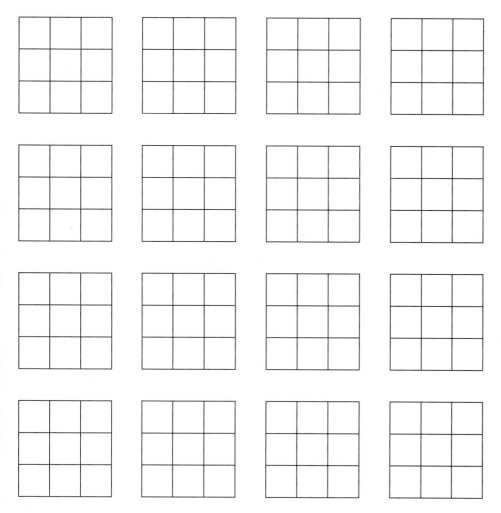

5 × 5 Magic Square Templates For EXPLORATION 1.5

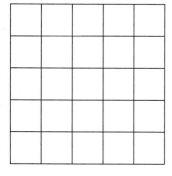

Looking Back on Chapter 1

1.   You have done several explorations during the course of this first chapter. Take a few minutes to think back on those explorations. What did you learn? Jot down your thoughts before reading on.

Now let us take some time to reflect on some specific things you have learned in this chapter.

2.   Select the one problem-solving tool on which you feel you have made the most progress during this chapter. You may pick one from the "4 Steps for Solving Problems" guidelines, or you may describe a tool you used that is not on that list.

   a.   Give the tool a name. If it is a tool mentioned in the book, you can use that name, but you are encouraged to develop names that describe your own perspective.
   b.   Show *how* you used that tool to solve a problem. Imagine describing this tool to a reader who does not yet have this tool in his or her repertoire. How can you help that reader to see not just *what* you did but also *how* this tool helped you to solve the problem so that the reader can get a sense of how this tool works?

3.   Many students enter this course not knowing how to check their answers, other than by checking their computation. What have you learned about being able to "verify and interpret results with respect to the original problem" (*Curriculum Standards*, p. 23) in the course of these explorations?

4.   We want students to be able to "justify their answers and solution processes" (*Curriculum Standards*, p. 29). Again, this idea is new for many students. In what ways has having to justify your answers and solution processes—to the instructor on an assignment, to a group member in a discussion, and to yourself while working on a problem—helped make you a stronger learner of mathematics?

5.   Describe any changes or shifts on any of your attitudes or beliefs about mathematics or about learning mathematics.

CHAPTER 2

Fundamental Concepts

The concepts of sets, functions, and numeration permeate elementary mathematics. The goals of the explorations in this chapter are for you to recognize and be comfortable with these fundamental concepts. Throughout the book, we will run into situations in which ideas related to sets and functions enable us to communicate more easily and help us to understand and solve problems. Similarly, a deeper understanding of numeration systems will enable you to develop what the NCTM calls mathematical power.

SECTION 2.1 **EXPLORING SETS**

As you will discover in your methods course, the concept of sets (and subsets) is an important aspect of the development of logical thinking in young children.

Do you know that Jean Piaget, a Swiss psychologist, used set concepts to help determine the extent to which a young child could reason abstractly? He learned how children's thinking developed by conducting "interviews" with children, one of which is illustrated below. A child is given a set of objects that are all mixed together. The child is then asked to separate the objects into two or three groups so that "each group is alike in some way." Do this yourself and record your different answers.

Children who have moved from what we call preoperational thinking to concrete operational thinking are able, fairly quickly and easily, to make subsets like the following: triangles, circles, and squares; big shapes and small shapes; white shapes and black shapes. Children who are still preoperational are not able to make these

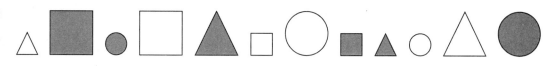

classifications as easily and consistently. It is not easy for them to hold the three attributes (color, shape, and size) of each object simultaneously in their minds. The following explorations are designed to enhance your ability to work with sets and subsets so that when you work with children, you will be able to recognize and nurture this important aspect of children's "mathematical" thinking.

EXPLORATION 2.1   **Understanding Venn Diagrams**

In this exploration, we will explore Venn diagrams as a tool for gathering and
reporting data about different subsets of a set.

Consider the following Venn diagram, in which the areas listed below represent
various sets of students:

- *U* represents the set of students in this class.

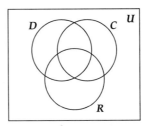

- *D* represents the set of students who have at
least one dog.
- *C* represents the set of students who have at
least one cat.
- *R* represents the set of students who have at
least one rodent, such as a rat, mouse, gerbil,
hamster, or guinea pig.

1.   Write your initials in the region in which your household belongs. If you live in a
dormitory, think of your parents' home or your previous home. Then discuss the
diagram with other members of your group. Each member should show where
his or her name was placed and why.

In Steps 2 through 5, refer to the class Venn diagram on an overhead projector or
blackboard.

2.   Which subset has more members: students who have cats and dogs, students who
have cats and rodents, or students who have dogs and rodents? How did you
figure this out?

3.   What fraction (percentage) of the class has at least one dog?

4.   How would you describe in words the subset of the class whose names are
outside all three circles?

5.   Describe a different subset of the class:

    a.   In words

    b.   With a Venn diagram

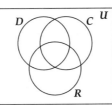

6.   How would you describe the subset of the class that has no pets?

    a.   Write down your thoughts about this question.

    b.   Discuss the question in your group or with the whole class. If you change
your mind as a result of the discussion, describe your new thoughts and
explain why you changed your mind.

EXPLORATION 2.2   **Gathering and Interpreting Data**

This exploration involves collecting data and then using set ideas to analyze that data.

Many writers have described our era as the Computer Age. Many items that are found in most homes today either did not exist 20 years ago or were found in only a small percentage of homes. Think of the household you belong to; if you live in a dormitory, think of your parents' home or your previous home. Does that household have a computer, an answering machine, or a compact disc player?

1.   *Ask the questions*   If we consider the students in this class as a subset of the set of college students, what questions might you ask concerning ownership of these items?

   a.   Write the questions that you would like to ask, and then discuss the questions.
   b.   After the class discussion, write the questions that will actually be asked.

2.   *Gather the data*   Determine how you will gather the data in order to answer the questions you are asking.

   a.   Write your proposed method.
   b.   After the class discussion, write the method that will be used. If this method is different from the one in part (a), discuss and then describe the advantages of the method that the whole class chose over the method that you described in part (a).

3.   *Answer the questions*

   a.   In your group, discuss how you can represent the information that you gathered in order to answer the questions.
   b.   Prepare a brief report that includes your answers to the questions, a description of how you determined the answers, and your representation of the information (a chart or diagram).

EXPLORATION 2.3   **Analyzing Different Representations**

Let's say you were going to gather information about the relative popularity of various recreational activities. For example, in the past month, how many students at your college have gone to a movie, a sporting event, or a cultural event?

1.   Decide on the number of categories and the description of each category. Then write down the questions that you want to be able to answer after collecting the information.
2.   Assume that someone has collected information that will enable you to answer your questions. Discuss the various ways in which you could represent this information. In the table on page 21, summarize the advantages and disadvantages or limitations of each kind of representation.

Looking Back

1.   You have done several explorations with sets. Take a few minutes to think back on those explorations. Select an important learning and describe that learning so that the reader gets a sense of what you learned, where and/or how that learning developed, and why it feels important.
2.   Describe a situation that knowledge acquired in these Explorations would help you to analyze.
3.   Describe something that you are still not clear about.

Table for   EXPLORATION 2.3, Analyzing Different Representations

Representation	Advantages	Disadvantages/limitations

SECTION   **2.2**   **EXPLORING ALGEBRAIC THINKING**

The concept of functions is one of the "big ideas" of mathematics. Although formal presentation of functions and other algebraic ideas does not happen until grade 8 or 9 in most schools, the foundation for understanding many fundamental algebraic ideas is laid in elementary school. An essential component of this work has to do with patterns. We see patterns almost everywhere, both in everyday life and in work situations, and children love playing with patterns. We connect this natural fascination with patterns to mathematical thinking by focusing on recognizing, describing, and extending patterns. In this way, children "develop the ability to classify and organize information" (*Curriculum Standards*, p. 60).

The notion of analyzing patterns and functional relationships has many real-life applications. One of the most poignant has to do with finding missing children. In one case, researchers helped investigators find two children who had been abducted by their father and had not been seen in eight years. Scott Barrows and Lewis Sadler developed a computer program that enabled them to take photographs of the five- and seven-year-old girls and create pictures of what they should look like eight years later. "The pictures showed up on TV one night and within 10 minutes a national hot line was getting calls from neighbors . . . by 7:30 the next morning the girls were in police custody."[1]

How had the researchers been able to do this? It seems that facial bones change in predictable ways throughout childhood. The two researchers determined relationships for 39 facial dimensions. By recognizing and analyzing the most important patterns, they were able to predict quite accurately what the girls would look like eight years later. In the explorations and investigations in this chapter, you will explore patterns and see how the concept of functions is related to understanding patterns.

[1] *Newsweek,* Feb. 13, 1989, p. 62.

EXPLORATION 2.4   **Relationships Between Variables**

This exploration is adapted from an article by Leah McCoy called "Algebra: Real-Life Investigations in a Lab Setting."[2] Although McCoy used these investigations with middle school students, many of them can be adapted for use in most elementary grades. On the web page for this book, you can find several different examples of these kinds of explorations at the elementary school level.

The key idea here is to collect data and then look for relationships between the two variables so that you can see a trend or pattern that enables you to predict the future! This is one of the primary reasons why functional relationships are so important. When the relationship between two variables is a functional one, we can predict the future. A simple function is the hourly wage. For example, if you are being paid at the rate of $8.00 per hour, you know that your earnings will be determined by the function:

Earnings = 8 times the number of hours you work

Not all functional relationships are this simple, though. I tell my students that although some aspects of mathematics are simple, many are complex—but that this is true also of most aspects of life, for example, marriage! Many students say they have math anxiety and avoid mathematics, but few people say they have marriage anxiety and avoid marriage—even if they "fail" once or twice—though some people do take some time off!

Your instructor will assign your group one of the questions below. Each group will perform the assigned exploration until you feel you have enough data to answer the question, and then each group will report its findings to the whole class.

The Explorations

1.   What is the relationship between the number of people and the time it takes to make a wave? Start with a group of five students and make a wave (as is done at sporting events). Collect more data with larger groups of people.

2.   What is the relationship between the number of copies needed and the time it takes to collate the copies? Your group will be asked to simulate a secretarial task in which the person must collate x copies of a 5 page document. For example, let's say you are to make 20 copies of a 5 page document. How long will it take to collate all 20 copies, including attaching a paper clip to each copy? (In actual situations, the copies would usually be stapled, but we don't want to ruin the blank sheets of paper.) Collect data with different numbers of copies, but keep the number of pages per document constant, i.e, 5. Otherwise, you will have a function with *two* variables.

3.   What is the relationship between the number of dominoes in a run and the time it takes for them all to fall? Make a domino run and determine how long it takes for all the dominoes to fall. Repeat the experiment with different numbers of dominoes. *Note:* In this case, you will keep the distance between dominoes constant.

4.   What is the relationship between the number of crackers eaten and the time, after eating them, before one can whistle? One person will be given x small crackers (such as oyster crackers or goldfish). How long after eating the crackers is it until the person can whistle? Repeat the experiment with different numbers of crackers.

[2] Leah McCoy, "Algebra: Real-Life Investigations in a Lab Setting." Mathematics Teaching in Middle School, 2(4) (February 1997), pp. 220–224.

5. What is the relationship between the height of an object and the length of its shadow? Go outside and measure the heights of various objects (including people) and the lengths of their shadows.

6. What is the relationship between the height of water in a beaker and the number of marbles in the beaker? You will be given a graduated beaker and a bunch of marbles. Fill the beaker partially full with water and record the height of the water. Add x marbles and note the increase in height. Repeat the experiment with different numbers of marbles.

7. What is the relationship between the number of drops of ink on a paper towel and the diameter of the circle formed by the ink? You will be given some paper towels, a bottle of ink, and a dropper. Carefully drop one or more drops of ink onto the paper towel and then measure the diameter of the circular region that is formed. Repeat the experiment with different numbers of drops.

8. What is the relationship between the height of a dot on a wall and the distance from a mirror to the observer? You will be given a mirror and bright dots that can be taped to a wall. Place the mirror on the floor close to a wall. Tape a dot to the wall. Stand next to the mirror and back away from the mirror until you can see the dot in the mirror. Repeat the experiment with the dot at different heights.

9. You will be given a ball and asked to collect data on two questions.

   a. What is the relationship between the number of bounces (n) and the height of the ball after the nth bounce? Repeat the experiment with as many bounces as you can accurately measure. Always drop the ball from the same height.

   b. What is the relationship between the number of bounces (n) and the time elapsed until the nth bounce? Repeat the experiment with as many bounces as you can accurately measure. Always drop the ball from the same height.

10. You will be given a number of jar lids and asked to collect data on two questions.

    a. What is the relationship between the diameter of a lid and the distance around the edge of the lid? Repeat the experiment with different lids.

    b. What is the relationship between the diameter of a lid and its area? Repeat the experiment with different lids.

The Procedures

Each group should go through the following steps in order to carry out its exploration.

1. Make sure everyone understands the problem.
2. Determine each member's role—recorder, measurer, and so on.
3. Discuss how the data will be measured and recorded—in inches, in centimeters, with rulers or stop watches, and so on.
4. If precision is an issue, do some practice runs to make sure you can record the data as precisely as needed.
5. Collect the data.
6. Determine your answers to the question and then make your report.

The Reports

Each group's report should include the following:

   a. A statement of your problem/question.
   b. A statement of what you did (your design).
   c. A chart or overhead containing the following information or an explanation of why some of the information is not relevant to your situation:

      ■ The independent variable (and unit)

- The dependent variable (and unit)
- A table of your data
- A graph
- An equation expressing the relationship between the variables
- The relationship between the variables, expressed in words

d. Your primary conclusions, including your degree of precision and your level of confidence in your conclusions. Describe any factors that limit your degree of precision and/or limit your ability to generalize.

e. A description of your biggest obstacle/problem and how you overcame/solved it.

EXPLORATION 2.5   **Growth Patterns**

Scientists, economists, and businesspeople are constantly confronted with measuring and predicting growth: How fast do certain bacteria grow? How fast is a disease (like AIDS) spreading? How fast is a company growing? How fast is a population (of a country or of an animal species) increasing? The conclusions they draw from the data affect decisions: whether to buy or sell the stock, to declare this disease a crisis, or to put this species on the endangered list. If they make the right decision, money is made or lives are saved; if they make the wrong decision, money or lives are lost.

Of course, elementary school children are not going to deal with such sophisticated problems, but they can explore growth situations at a level that develops their mathematical abilities in the five process standards—problem solving, reasoning, communication, making connections, and representation. The following explorations have been designed to challenge you and help you grow, and they can be done, modified according to the age of the children, with elementary school children. Variations of these explorations are found in many elementary school textbooks and curriculum materials (see references on the web site).

PART 1: Figurate numbers

Humans have long been fascinated by number patterns. The ancient Greeks thought of numbers as quantities made up of units, which they often expressed as dots.

Therefore, it was natural for them to "see" numbers—and number patterns—geometrically. Thus, they spoke of triangular numbers, square numbers, and so on. The first few triangular numbers and square numbers are shown below.

Exploration of these kinds of numbers, called **figurate numbers,** is done in many elementary schools. One reason is that it is developmentally appropriate for children, who generally need more concrete representations of mathematical ideas. Another reason is that there are many opportunities for doing good mathematics with these numbers—looking for patterns, making and testing predictions, and learning how to develop formulas from patterns (remember Gauss's formula from Investigation 1.5: The Sum of the First 100 Numbers).

Trinumbers

1.   Let us first examine the set of trinumbers, which are triangular numbers that have been "hollowed out." The first six trinumbers are shown following. The question you will be asked to answer is "How many dots will be in the *n*th trinumber?"

However, we will build up to that question. For many students in this course, this is a huge jump. Thus, we will use the analogy of steps. Rather than taking one big step, we will take a number of smaller steps.

a. First, write below each trinumber the number of dots it takes to make that trinumber. How did you count those dots? If you simply counted 1, 2, 3, . . . , then you are not likely to notice patterns. As was presented in Chapter 1 and reinforced in Chapter 2, looking for and making sense of patterns is a key part of "doing mathematics." Now that you know how many dots it takes to make each of the first six trinumbers, what observations do you make, or patterns do you see, that might help you to answer the question?

b. Your instructor will now have you share your observations with your partners or the whole class.

c. Some of these observations will be more useful than others in helping you to find the expression for the nth trinumber. Working at first on your own and then with your partners, try to make use of observations and patterns to determine the number of dots in the nth trinumber. It may be helpful to make a table.

Square Numbers

2. Let us look now at the set of square numbers. The first five square numbers are shown on page 27.

a. How many dots does the sixth square number contain?

b. How many dots does the tenth square number contain?

c. How many dots does the nth square number contain?

Triangular Numbers

Most students find it relatively easy to answer the three questions above, because the pattern is relatively simple. Let us now explore the set of triangular numbers. The Greeks saw the set of triangular numbers as literally *emerging* from the counting numbers. That is, they created them in the following manner: Put one dot on top of two dots; this makes a triangle. Now put this triangle on top of three dots; this makes a new triangle. Continue in this manner. Because the stacks (except for the first one) look like triangles, the Greeks called this set of numbers triangular numbers. Note that the Greeks called 1 the first triangular number.

3.   **a.** How many dots does the fifth triangular number contain? Write *how* you determined this amount.

b. How many dots does the sixth triangular number contain? Write *how* you determined this amount.

c. What patterns do you see that would enable you to predict the number of dots in the next triangular number? Write down the patterns you see and then compare your observations with those of others in your group. Add to your list useful patterns that others observed.

4. **a.** How many dots does the *n*th triangular number contain? Take a few minutes to explore this question on your own. Show your thinking. If you are stuck, consult the "4 Steps for Solving Problems" guidelines.

   b. Compare your ideas with those of other members of your group. Add any new ideas to your list.

   c. After the discussion, state the value of the *n*th triangular number and explain how you arrived at that value. If you are not sure, summarize your current thinking and where you are stuck.

Once we have a way of describing the *n*th triangular number, we have uncovered the functional relationship. That is, we can now pick any triangular number and tell how many dots that triangular number contains, whether it is the 10th or the 250th triangular number. This leads into the next part of the exploration: how to communicate our discoveries and observations.

Relationships Between Figurate Numbers

There are many ways to look at the relationship between square numbers and triangular numbers and many ways to describe what we see.

5. *Interpreting a new notion*   The following formula expresses *one* way in which the two sets of numbers are related:

$$S_n = T_{n-1} + T_n$$

   a. Translate this equation into English.

   b. Exchange your translation with a partner and examine carefully your partner's translation with respect to accuracy and clarity. Discuss each of your translations until you both feel you understand the formula and can describe the formula in words.

   c. Work with your partner to justify this relationship—that is, to explain why it is true. Write your first draft of a justification.

6. *Translating words into notation*   The following sentence describes, in words, another way in which two sets of numbers are related: The square of any odd number is one more than eight times a specific triangular number. First, make sure you understand this statement. For example, 5 is an odd number, and $5^2 = 25$; 3 is a triangular number, and $3 \cdot 8 + 1 = 5^2$.

   a. Translate this sentence into notation.

   b. Exchange your notation with a partner and critique your partner's notation with respect to accuracy and clarity.

Other Figurate Numbers

7. Look at the pentagonal numbers below.

   a. Describe patterns that you observe. You may want to draw the next few pentagonal numbers.

b. Write questions or hypotheses about the pentagonal numbers and their relationship to triangular and square numbers.

8. Look at the hexagonal numbers below.

a. Describe patterns that you observe. You may want to draw the next few hexagonal numbers.

b. Write questions or hypotheses about the hexagonal numbers and their relationship to triangular and square numbers.

9. Make a table like the one shown below. Write in your expressions for the value of the nth triangular number and the nth square number. Determine the value of the nth pentagonal number and the nth hexagonal number. Show your work.

Type of number	Value of the nth term
Triangular	
Square	
Pentagonal	
Hexagonal	

10. Certain relationships among different figurate numbers enable us to predict the value of the nth number for *any* figurate-number family, such as the nth octagonal number. Describe at least one relationship among the formulas in Step 9 that would enable you to make such a prediction. Explain.

PART 2: Squares around the border

The following problem is a popular growth pattern problem in elementary schools because it is so rich, as you will soon find out. In fact, the February 1997 issue of *Teaching Children Mathematics* has a whole article describing how this problem can be adapted for grades K–2, for grades 3–4, and for grades 5–6. I encourage you to look it up in your library. I have also seen variations of this problem at several conferences and in other articles. With young children, the problem is often presented with blue and white tiles (that you can get at a tile store). The blue tiles (in the center) represent the water in the pool, and the white tiles represent the border around the pool.

1. Below you can see the first five figures in this growth pattern. Count the number of squares in the border of each figure, and then, from the patterns you see and the observations you make, determine the number of squares around the border of the *n*th figure. As in Part 1, we will break the problem into a series of smaller steps.

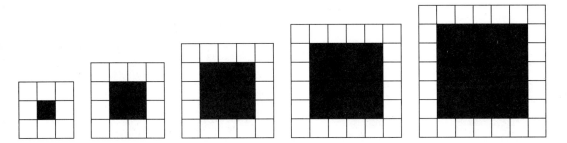

   a. How many squares are there in the border of the first figure? of the second figure? of the third figure? of the fourth figure? What patterns do you see, in the figure and/or from how you counted the number of squares? Draw additional figures if you feel that would help.
   b. Your instructor will now have you share your observations with your partners or the whole class.
   c. Working at first on your own and then with your partners, try to make use of observations and patterns to determine the number of squares in the border of the *n*th figure.

2. One of the themes of the book is the notion of multiple representations. Often, we see new aspects of a situation when we examine it from another perspective. We will do that here—the other perspective is to see what graphs tell us about the relationships among the figures, the number of squares around the border, and the number of squares in the middle.

   a. Fill in the table on page 33 and then make two graphs. In the first graph, the independent variable will be the number of the figure (e.g., 1st, 2nd, 3rd, etc.), and the dependent variable will be the number of squares around the border of that figure. In the second graph, the independent variable will again be the number of the figure, but the dependent variable will be the number of squares in the middle of that figure.
   b. Record any observations you make from the two graphs.
   c. How would you describe, in words, the growth of the number of squares around the border and the growth of the number of squares in the middle?

Extensions

As you are finding, many rich problems have more to offer even after you have found the answer. They are extendible! Let us examine three extensions of the squares-around-the-border problem.

3. **a.** Determine the fraction of the total area that is contained by the "pool" (inner shaded region) for the first five figures.
    b. Describe the changes you see in the fraction.
    c. What fraction will be pool in the nth figure?
    d. Make a graph of this situation: The independent variable is the number of the figure, and the dependent variable is the fraction of the area that is pool. Describe this function in words.

4. The growth pattern shown below adds one more layer to this problem. We have the white outer border and a shaded inner border. How many squares would make up the nth outer border (white squares)?

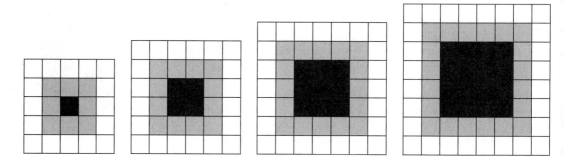

5. This extension comes from realizing that we don't have to have a square "pool" and a square "border." How many squares are in the border of the nth figure for the growth pattern shown below?

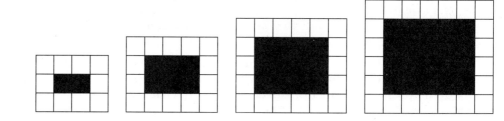

Table for EXPLORATION 2.5, PART 2, Step 2(a)

Figure number	Number of squares around the border	Number of squares in the middle	Notes on how you determined this
1			
2			
3			
4			
5			
6			
7			

EXPLORATION 2.6   **Another Growth Pattern**

Pattern blocks can be found in most elementary classrooms. They are one of the most versatile of the mathematics manipulatives because they can be used to explore mathematical concepts from functions to fractions to geometry to measurement.

A common pattern that has applications in many fields is what we call a growth pattern. Biologists are interested in understanding how fast bacteria or animals or plants are increasing. Epidemiologists want to know how fast an epidemic spreads. Businesspeople want to know how fast their company is growing. Employees want to know how fast their retirement fund is growing. We will use pattern blocks to model a simple growth situation. Note that I am using the word *simple* here to mean "not complex," as opposed to "not difficult."

1.   Take a few minutes to play with the pattern blocks. Note your observations and then share these observations with other members of your group.

Building a Pattern Block Patio

Let's say that an eccentric couple has decided to make a huge patio in their back yard with concrete blocks that are in the shape of giant pattern blocks. The pattern they want to create will consist of rings of hexagons. That is, they begin with a hexagon and then make bigger and bigger rings around this hexagon. The diagram at the left shows the initial block, and the diagram

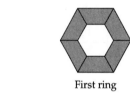

Initial block   First ring

at the right shows the first ring. If the hexagon costs $6 and the other blocks are valued proportionately, by surface area, then the initial block costs $6 and the first ring costs $18.

The ultimate question is how much the *n*th ring will cost. That is, can you find patterns in this growth so that if the couple were to ask how much the 10th ring would cost, you could tell them without having to make a physical model of the whole design? What about the 15th ring? the 20th ring? The patterns and the ability to make predictions from the observed patterns will emerge if you begin with the initial block and then build from there.

2.   **a.**   Build the second and third rings and determine their respective costs.
     b.   What patterns do you notice—either patterns that enable you to make the third ring or patterns in the numbers for the cost of the first, second, and third rings? Write down your observations.
     c.   Do you see any patterns that might enable you to predict the cost of the fourth ring? If possible, make a hypothesis before you build the fourth ring.
3.   **a.**   Build the fourth ring and determine its cost.
     b.   What patterns do you notice—either patterns that enable you to make the fourth ring or patterns in the numbers for the cost of each of the rings? Write down your observations.
     c.   Do you see any patterns that might enable you to predict the cost of the fifth ring? If possible, make a hypothesis before you build the fifth ring.
4.   **a.**   Build the fifth ring and determine its cost.

    b. What patterns do you notice—either patterns that enable you to make the fifth ring or patterns in the numbers for the cost of each of the rings? Write down your observations.

    c. Do you see any patterns that might enable you to predict the cost of the sixth ring?

5. How much does the nth ring cost? If you have difficulty with this question, and many students do, the following ideas are often helpful.

- Make a table like the one shown and fill in the Cost column.

- Think back to Chapter 1 when we expanded our tables in Investigation 1.1 (Pigs and Chickens) and in Investigation 1.5 (The Sum of the First 100 Numbers). Would an additional column help you to predict the cost of the nth ring?

Ring	Cost	
0	6	
1	18	
2		
3		
4		
5		
n		

6. Prepare a brief presentation of your group's work. One member of your group will present your answer to the question and explain how your group arrived at this answer.

7. What did you learn from other groups' presentations and from the class discussion that followed?

Looking Back on Exploration 2.6

Describe one way in which your ability to recognize and extend patterns increased as a result of this exploration.

SECTION ◆ **2.3** **EXPLORING NUMERATION**

One day an elementary teacher, Georges Ifrah, was asked by one of his students, "How did numbers start? When did people learn to count?" At the time, all he could say was "I don't know" and "a long time ago." To answer those questions, he did quite a bit of research and wrote a fascinating book called *From One to Zero: A Universal History of Numbers*. Historical records show that many different kinds of number systems were created as humans' understanding of mathematics increased. You will spend some time creating your own numeration system so that you will be able to understand more deeply how we use numbers.

EXPLORATION 2.7 **Alphabitia**

Imagine that you are a member of a small tribe that lived thousands of years ago, when people were making the transition from being hunter-gatherers to becoming farmers. You have a numeration system that is alphabetically based, so you are called Alphabitians. Like many other ancient peoples your numeration system is finite. For any amount greater than Z, you have no symbol; you just call that amount "many."

Amount	•	• •	•••(C)	••••(D)	•••••(E)		
Alphabitian numeral	A	B	C	D	E	...Z	many

PART 1: Inventing a new system

Now that your tribe has settled down, you have "many" sheep and "many" ears of corn. Without an adequate numeration system, figuring out how many more sheep you have this year than last year and determining each family's share of the corn harvest is very tedious. A young woman in your tribe has excitedly announced that she has invented a new counting system with which she can represent any amount using only the symbols A, B, C, D, and a new symbol she calls zero and writes as 0. Unfortunately for your tribe, this young woman died on a hunting trip. However, she left behind some artifacts that she was going to use to help you learn the new system. These artifacts are called flats, longs, and units.

flats   longs   units

Because the visionary member of your tribe is no longer with you, it is up to you to invent the new numeration system that your tribe desperately needs. That is, you need to develop a system that lets you represent *any* amount using only the symbols A, B, C, D, and 0.

1. Cut out a number of flats, longs, and units from graph paper or use the manipulatives provided by your instructor.
2. Take some time to sit down with your partners and create a numeration system using only the symbols A, B, C, D, and 0.
3. Some groups will find that one member proposes a system that makes sense to everyone quickly. Other groups will discuss and debate two or more systems

before finally deciding on one system. After you have explored different alternatives, answer the following questions before proceeding to Part 2.

  a.   Does your system make sense to every member of the group?
  b.   Imagine that you will be explaining your system to the council of the elders. How would you explain your system to them?
  c.   If your group decided among two or more possibilities, what made you choose one system over the other(s)?

PART 2: Communicating different systems

At a time specified by your instructor, each group will post its system on the wall, and each student will have time to look at the other systems.

Old	New	Picture
A		.
B		:
C		⋮
D		⋮
E		I
F		I˙
...		
X		
Y		
Z		

1.   Make a group poster. Each group's poster needs to contain the following:

  ■ The name of your group (feel free to be creative and playful).

  ■ The table on page 41, showing your symbols for at least the first Z numbers.

  ■ Directions that will help other students to understand the system *(Remember: You will not be there to explain it to other students.)*

  ■ A frank and honest assessment of the advantages and the disadvantages or limitations of your new system

2.   *First discussion of the systems: observations and questions*   After students have had a chance to examine the different systems, there will be a class discussion in which you will be asked to share your observations and questions about other systems.

  a.   Jot down observations or questions that you would like to remember.
  b.   What did you learn from this discussion?

3.   *Taking different systems for a test drive*   After examining all the systems, select the two systems that you like best and take them for a "test drive." You may choose the system your group developed as one of the two systems. Each test drive must include the following tasks:

  a.   Try counting up to double Z (that is, the amount equivalent to Z + Z in the original Alphabitian system). Record your symbols and describe any difficulties you had and any uncertainties you have. For example, you may see two reasonable possibilities for a certain amount. Reminder: The goal here is "sense making" as opposed to "just do it." To the extent that you get into this role play, to that extent you will better understand the difficulties young children have when learning to count in our numeration system!

  b.   Make up and solve some simple story problems involving addition (e.g., new sheep born), subtraction (e.g., selling sheep), multiplication (e.g., planting rows of trees), and division (e.g., dividing ears of corn among several families). Please stay in your role as Alphabitians. Among other things, this means forgetting the computation procedures you know as Americans. Thus, when you divide, think how you (as an Alphabitian) might determine the answer to the question in your story problem.

   c. Evaluate each system. First, summarize its strengths and weaknesses or limitations. Then make any recommendations for improving the system.

4. *Second discussion of systems: choosing one system*   At this time, your class will decide upon one system that you will all learn and use. You might select one of the systems created, or you might combine two systems (each with certain advantages) and then select the combined system.

   a. As different students nominate a system and then debate the advantages and limitations of that system, make notes about any points that you want to remember.
   b. What did you learn from this discussion?

PART 3: Learning about the new system

1. **a.** After the class selects the system that it will use, take time to learn that system. You may want to review the "4 Steps for Solving Problems" before you begin.
   b. Once your group feels that each member understands the new system, take a few moments to jot down structures of this new system that help you to count quickly and confidently.

   For Steps 2 and 3, take some time to answer the following questions. Answer them using manipulatives, diagrams, or symbols.

2. What number comes after each of the following? Briefly explain how you answered each question.
   a. AA   **b.** BD
   c. BAD   **d.** ABD

3. What number comes before each of the following? Briefly explain how you answered each question.
   a. D0   **b.** B00   **c.** D0D0

4. Compare your responses to Steps 2 and 3 with those of other members of your group.

   Take some time to analyze mistakes. Were they just careless, or do they point to some aspects of this system about which you are still fuzzy?

   Your instructor may have you do one or more additional worksheets.

Looking Back on Exploration 2.7

1. Describe the most important learnings from this exploration. In each case, first describe what it was that you learned and then describe how you learned it.
2. Describe something in this exploration that you are still not clear about.
3. Examine the five process standards: Problem Solving, Reasoning, Communication, Connections, and Representation. Select one of the subheadings from this set that you feel you "own" more as a result of this exploration. Describe what you learned.

Alphabitia Table for EXPLORATION 2.7

Old symbol	New symbol(s)	Picture of what the amount looks like	Explanation
A		·	
B		:	
C		⋮	
D		⋮	
E		|	
F		|·	
G		|⋮	
H		|:	
I		|⋮	
J		||	
K		||·	
L		||⋮	
M		||:	
N		||⋮	
O		|||	
P		|||·	
Q		|||⋮	
R		|||:	
S		|||⋮	
T		||||	
U		||||·	
V		||||⋮	
W		||||:	
X		||||⋮	
Y		▦	
Z		▦·	

EXPLORATION 2.8   **Different Bases**

In this exploration, we will explore several different bases and their relationships to one another. The primary purpose of these explorations is to deepen your understanding of base 10 — that is, your understanding of place value and the function of zero. I have found that the structures of a system are often best seen by putting the familiar in an unfamiliar context; this was the reason for the Alphabitian exploration. Over the past ten years, I have observed that an increasing number of elementary teachers have their students explore different bases. These teachers have told me that the explorations with different bases help their students come to a better understanding of place value and how base 10 works, and that this knowledge results in better problem-solving in base 10.

We will begin by exploring three different bases after noting an important caution: Working in different bases is often difficult at first. If you find yourself groping about, consider the following tools to help you:

Connections

Can I connect what I know about base 10 to this new base?

Problem-Solving Strategies

- What if I made a table? Would that help?
- As I make my tables, what patterns do I see?

Transition numbers are often difficult; in base 10, for example, 99 is a transition number. When you encounter transition numbers, consider the following questions:

- Might it help to use manipulatives?
- Might it help to stop and say what the symbols mean?

Patterns

- Are there patterns that are true for all bases?
- Are there patterns that are true for some bases but not others?

You may have already surmised that the new Alphabitian system is actually a base 5 system that uses A, B, C, and D instead of 1, 2, 3, and 4. Before proceeding with the following explorations, you might want to make a base 5 counting chart (1, 2, 3, 4, 10, 11, 12, 13, 14, 20, 21 . . .) and become comfortable with this system.

PART 1: Learning about different bases

Base 6

The following activities are designed to help you understand base 6.

1. a. If your instructor has base 6 manipulatives, get them if you wish. If your instructor does not have them, you might want to make your own, using graph paper.
    b. Take some time, individually or in a group, to learn how to count in base 6.
    c. As you learn about this new base, remember to have a group check-in frequently to be sure you are not leaving anyone behind. One way to do this is to ask a question and see whether everyone can answer it and explain it.

For example, pick a number and ask what the next number is or what the previous number is. Record your work.

2.   Once you have determined that all members of the group understand base 6, respond to the following questions.

   a.   What helped you to accomplish this task?
   b.   What patterns do you see in this base?

3.   Now that you are able to count in base 6, try the following exercises, both to assess your understanding and to stretch your understanding. Answers are on page 48.

   a.   What comes after each of the following?[3] 25_6 555_6 1235_6
   b.   What comes before each of the following? 40_6 300_6 12340_6

4.   Describe one area in which you had initial difficulty—for example, deciding what comes after 555_6. Describe the difficulty; for example, did you have no idea, or were you debating between two possible answers? Describe what it was that helped you get unstuck.

Base 2

The following activities are designed to help you understand base 2.

5.   **a.**   If your instructor has base 2 manipulatives, get them if you wish. If your instructor does not have them, you might want to make your own, using graph paper.
   b.   Take some time, individually or in a group, to learn how to count in base 2.
   c.   As you learn about base 2, remember to have a group check-in frequently, as in Step 1(c).

6.   Once you have determined that all members of the group understand base 2, respond to the following questions:

   a.   What helped you to accomplish this task?
   b.   What patterns do you see in this base?

7.   Now that you are able to count in base 2, try the following exercises, both to assess your understanding and to stretch your understanding. Answers are on page 48.

   a.   What comes after each of the following? 110_2 1011_2 101111_2
   b.   What comes before each of the following? 100_2 111_2 1010_2

8.   Describe one area in which you had initial difficulty—for example, deciding what comes after 1011_2. Describe the difficulty; for example, did you have no idea, or were you debating between two possible answers? Describe what it was that helped you get unstuck.

9.   Some students find base 2 the hardest to learn, whereas others find it the easiest. Why do you think this is so?

Base 12

The following activities are designed to help you learn about base 12.

---

[3]A subscript at the end of a number indicates the base. Thus, 25_6 means the amount 25 in base 6. When there is no subscript, the number is in base 10.

10. **a.** If your instructor has base 12 manipulatives, get them if you wish. If your instructor does not have them, you might want to make your own, using graph paper.

    b. Take some time, individually or in a group, to learn how to count in base 12. Then read part (c).

    c. Base 12 presents a problem that did not come up in the other bases: How do you represent the amount that in base 10 is called "10" in base 12? What do you think? Write down your first thoughts.

    d. Discuss the problem in part (c) in your group. Discuss why this is a problem, and discuss possible ways to resolve it. Write down your thoughts.

11. **a.** Once you have resolved the problem of new digits by consulting your instructor or from a class discussion, take some time, individually or as a group, to learn how to count in base 12.

    b. As you learn about base 12, remember to have a group check-in frequently, as in Step 1(c).

12. Once you have determined that all members of the group understand base 12, respond to the following questions:

    a. What helped you to accomplish this task?

    b. What patterns do you see in this base?

13. Now that you are able to count in base 12, try the following exercises, both to assess your understanding and to stretch your understanding. (*Note*: There is no answer key for this question because it depends on the symbols that your class invents for the two numbers after 9 in base 12.)

    a. What comes after each of the following? 39_{12} 79_{12} 909_{12}

    b. What comes before each of the following? 100_{12} 450_{12} 1010_{12}

14. Describe one area in which you had initial difficulty — for example, deciding what comes after 19_{12}. Describe the difficulty; for example, did you have no idea, or were you debating between two possible answers? Describe what it was that helped you get unstuck.

PART 2: Similarities and differences among the bases

At this point, you can count in several different bases. In fact, if you stop to think for a moment, you could count in base 3 without further instruction, as long as you understand what it means to say base 3.

1. Take a minute to see if, indeed, you can count in base 3. If you can't, your understanding of bases may be primarily procedural, or "rented."

When we examine bases, we find that certain rules or patterns are true for any base, whereas other patterns occur in some, but not all, bases. This observation is the focus of this exploration.

2. Take some time to look at the different bases. What patterns do you see in all bases? What patterns do you see in some bases but not in others? Record your observations in the table on page 49. Also note what sparked the discovery — perhaps looking at a table or hearing a comment by someone else.

3. Listen to other students describe their patterns. Write down the new patterns that you heard.

Looking Back

Now that you have learned to count in many bases (you are multinumerate, as opposed to multilingual), take some time to reflect on what you have learned and then answer the following questions.

1.   To what extent did this exploration deepen your understanding of base 10?
2.   Describe one problem-solving tool that was particularly useful in these explorations. Your description needs to be specific enough that the reader can see how that tool actually changed your understanding.
3.   What was the most important thing you learned from investigating different bases?

PART 3: Translating from another base into base 10

A traveler going from the United States to Great Britain has to convert dollars to pounds. Similarly, we will now investigate the process of converting amounts from one base to another. Before we proceed, I need to emphasize strongly that the point of the following explorations is not to master the procedure for translating from one base to another. Rather, these explorations have two aims: (1) to give you practice in applying the problem-solving tools you are developing, and (2) to deepen your understanding of base 10, which is the base that you will (at least implicitly) teach your future students, whether you are a kindergarten teacher helping your students learn to count, a second-grade teacher helping your students learn how to subtract, or a sixth-grade teacher helping your students learn decimals.

To help keep you focused on this goal and to make the process a bit more fun, let us go into the future and imagine that there are intelligent beings on every planet of the solar system. The beings on Mercury use what we call base 2; the beings on Venus use what we call base 5; the beings on Mars use what we call base 6; and the beings on Jupiter use what we call base 12.

1.   Each planet would say that *it* has base "10." Do you see why? Discuss this question in your group, and then write down why each planet would say it has base "10." Imagine explaining this to a classmate who was not present for this discussion and whose initial reaction to the statement is a puzzled expression.
2.   The United Nations of the Solar System is having its monthly interplanetary meeting. You are the hotel clerk who is handling reservations. Unfortunately, the faxes that come from each planet contain numbers in that planet's own base. Thus, you have to determine how many rooms to reserve for each delegation.
   a.   The Mercury delegation says that it needs 1101 rooms. How many rooms will you reserve?
   b.   The Venus delegation says that it needs 34 rooms. How many rooms will you reserve?
   c.   The Mars delegation says that it needs 23 rooms. How many rooms will you reserve?
   d.   The Jupiter delegation says that it needs 18 rooms. How many rooms will you reserve?
3.   Think about how you translated from each of these bases into base 10. Develop an algorithm that can be used for each of the bases, using the following guidelines. Imagine that you are going home soon and that the night clerk is not very confident working with different bases. Write directions for this clerk so that she can make the appropriate number of reservations, regardless of which base is used. At the same time, imagine that the night clerk is curious and wants to

understand. Draw a line down the middle of the page. On the left, describe what to do. On the right, explain why this works; in other words, justify each step in the algorithm.

4. Now imagine that the United Nations of the Solar System is having its biannual interplanetary conference. The number of participants at this conference is much greater than at the monthly meeting. As before, imagine that you are the clerk and need to reserve the appropriate number of rooms.

   a. The Mercury delegation says that it needs 100110 rooms. How many rooms will you reserve?
   b. The Venus delegation says that it needs 403 rooms. How many rooms will you reserve?
   c. The Mars delegation says that it needs 203 rooms. How many rooms will you reserve?
   d. The Jupiter delegation says that it needs 109 rooms. How many rooms will you reserve?

5. Once your group is confident that you have reserved the correct number of rooms, take a moment to look back on what you have learned. Did your algorithm from Step 3 work for larger numbers? That is, did it stretch well or did it break down? If it did not stretch well, explain the new algorithm that you developed.

PART 4: Translating from base 10 into other bases

1. The United Nations of Earth has held new elections, and it is time to send the new delegations to each of the planets. You are in charge of sending a fax to each planet telling them how many people are in the new delegation. However, you realize that the people who receive the messages may not translate accurately from our base into their base. You have decided that you will tell them the number of people in the delegation in their system. Below are the numbers of people in the delegation to each planet in base 10.

   a. The delegation to Mercury will consist of 34 people. Translate this amount into the base used on Mercury.
   b. The delegation to Venus will consist of 34 people. Translate this amount into the base used on Venus.
   c. The delegation to Mars will consist of 34 people. Translate this amount into the base used on Mars.
   d. The delegation to Jupiter will consist of 34 people. Translate this amount into the base used on Jupiter.

2. Think about how you translated from base 10 into each of these bases. Develop an algorithm that can be used for each of the bases, using the following guidelines. Imagine that you need to give directions to the person who will send faxes from the United Nations of Earth to each of the planets after the next elections. Draw a line down the middle of the page. On the left, describe what to do. On the right, explain why that works; in other words, justify each step in the algorithm.

3. Now imagine that you are a travel agent on Earth handling reservations for tours to each of these planets. You need to reserve the appropriate number of rooms.

   a. The tour to Mercury has 208 people. Translate this amount into the base used on Mercury.
   b. The tour to Venus has 208 people. Translate this amount into the base used on Venus.
   c. The tour to Mars has 208 people. Translate this amount into the base used on Mars.

   d. The tour to Jupiter has 208 people. Translate this amount into the base used on Jupiter.

4. Once your group is confident that you have reserved the correct number of rooms, take a moment to look back on what you have learned. Did your algorithm from Step 2 work for larger numbers? That is, did it stretch well, or did it break down?

Looking Back on Exploration 2.8

1. Describe the most important things that you learned while translating to and from base 10. In each case, first describe what it was that you learned and then describe how you learned it.
2. Describe something from this exploration that you are still not clear about.
3. Examine the five process standards: Problem Solving, Reasoning, Communication, Connections, and Representation. Select one of the subheadings from this set that you feel you "own" better as a result of this exploration. Describe what you learned.

Answers to Exploration questions

Exploration 2.8: Base 6, p. 44

30_6 comes after 25_6. 1000_6 comes after 555_6. 1240_6 comes after 1235_6.
35_6 comes before 40_6. 255_6 comes before 300_6. 12335_6 comes before 12340_6.

Exploration 2.8: Base 2, p. 44

111_2 comes after 110_2. 1100_2 comes after 1011_2. 110000_2 comes after 101111_2.
11_2 comes before 100_2. 110_2 comes before 111_2. 1001_2 comes before 1010_2.

Table for EXPLORATION 2.8,   **PART 2, Step 2**

Patterns common to all bases	What sparked this discovery?

Patterns common to certain bases (specify the bases)	What sparked this discovery?

EXPLORATION 2.9   **A Place Value Game**

Materials

- A six-sided die in which the 6 is covered with a zero
- The game boards provided on page 53. The grids have spaces for a 4-digit number, a 3-digit number, a 2-digit number, and a 1-digit number.

Rules for competitive version

1. Each player has a separate game board.
2. After each roll of the die, players decide where on the grid to write the number.
3. After ten rolls, each player determines the sum of the four numbers—that is, the 4-digit, 3-digit, 2-digit, and 1-digit numbers. The person with the largest sum wins.

1. Play the first game as follows:
   a. Do not speak during the game.
   b. In the blanks provided on the game board, record the order of the ten throws of your die. For example, 5 3 4 0 2 4 5 2 3 4.
   c. After the first game, take some time to think about your choices. Also think about whether you learned something from playing the game or from watching the other player. Can you express what you learned in a way that would make sense to someone who has not played the game? Write down the strategies that you learned from the first game.

2. Play the second game as follows:
   a. Record the order of the ten throws of your die.
   b. Every time you write a number in the grid, explain why you chose that location.
   c. After the game, reflect on your strategy. Describe the strategies you learned from playing the second game or from watching the other player.

Looking Back on Exploration 2.9

1. Summarize your strategies for this game, and briefly justify them.
2. What did you learn from playing this game?

Game boards for   EXPLORATION 2.9

1. — — — — — — — — — —

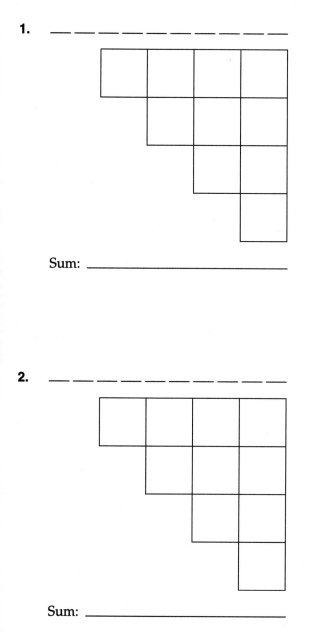

Sum: _____

2. — — — — — — — — — —

Sum: _____

The Four Fundamental Operations of Arithmetic

In Chapter 2, you explored three fundamental concepts of the elementary school curriculum: sets, function, and numeration. As Alphabitians you constructed a numeration system that is much more powerful than early numeration systems, and you gained a deeper understanding of the concepts of base, place value, and the role of zero in this system. This chapter builds on that knowledge so that you can see how our numeration system is connected to the fundamental operations of addition, subtraction, multiplication, and division. Understanding these connections will help you develop "operation sense," which in turn will help you with mental arithmetic—doing computations in your head—and estimation.

SECTION 3.1 EXPLORING ADDITION AND SUBTRACTION

Can you remember learning to add and subtract in first and second grade? As adults you probably don't think twice about adding and subtracting, but these operations are often difficult for many youngsters, especially when regrouping is involved. One way to help you better understand what these operations mean is to have you do problems in a different setting.

In Chapter 2 you put yourself in the role of a member of the Alphabitian tribe (Exploration 2.7), and in that role you created a numeration system. You will be asked to remain Alphabitians for a while longer and to add and subtract in that system.

EXPLORATION 3.1   **Inventing and Understanding Addition in Alphabitia**

1.   Make up and solve at least two addition story problems in Alphabitia and then discuss in your group how you solved them. Respond to the following questions:

   a.   What knowledge of counting, place value, and bases did you use and need to know?

   b.   If you encounter difficulty, what knowledge or tool(s) got you past the difficulty?

2.   One context for addition is to combine two different sets (Raga has BC sheep and Maru has BD sheep; how many do they have altogether?). What other contexts for addition do you find in your story problems?

3.   Now let us explore and understand how we might find the sums when the numbers are larger. It may surprise you to learn that the algorithms for addition, subtraction, multiplication, and division that you learned in elementary school are not the only possible algorithms. Over the centuries, many different algorithms have been invented, and even today, children in some countries and some parts of the United States learn different algorithms. You will investigate some of these other algorithms later in this chapter.

   A major goal in the explorations in this section and the next section is for you to be able to determine sums, differences, products, and quotients in ways that make sense to you and that you can justify (that is, you can explain why they work). In so doing, some of you will actually reinvent algorithms that were invented by other people—or you may even invent a totally new algorithm!

   a.   Now try these Alphabitian problems:

$$
\begin{array}{r} BB \\ +CC \\ \hline \end{array}
\qquad
\begin{array}{r} BC \\ +CD \\ \hline \end{array}
$$

   b.   What knowledge of counting, place value, and bases did you use and need to know?

   c.   If you encountered difficulty, what knowledge or tool(s) got you past the difficulty?

4.   Now let us examine some addition problems with larger numbers. Focus on making sense of the process, not simply on "getting" the procedure.

   a.   Do the following problems according to your instructor's directions.

$$
\begin{array}{r} BCD \\ +\ DB \\ \hline \end{array}
\qquad
\begin{array}{r} B0B \\ +ABC \\ \hline \end{array}
\qquad
\begin{array}{r} C0D \\ +CAB \\ \hline \end{array}
\qquad
\begin{array}{r} D0D0 \\ +BABA \\ \hline \end{array}
$$

   b.   Compare your *solution path* with those of other students in your group and explain your process. Your goal in this discussion is to understand how each group member made sense of the problem in order to find the sum successfully .

   c.   After you have heard about each person's solution path, stop and reflect on what differences you note. For example, did some use manipulatives or solve the problems using diagrams, and did some do the problems without manipulatives or diagrams? Did some start at the left and work to the right, and did some start at the right (ones place) and move to the left? Note any differences that stand out.

EXPLORATION 3.2   **Exploring Patterns in Alphabitian Addition**

Being able to add large numbers easily not only requires an understanding of how our numeration system works but also requires that the "addition facts" be internalized. For example, if you have to stop and figure out what B + C is when adding ACDB + B0CC, the process will be slow. Many young children simply learn addition facts by rote. However, there are many, many patterns in mathematics, and these patterns can also be used to help children learn addition facts. One important exploration that will help you internalize addition facts is to make an addition table and then recognize and analyze patterns.

1. *Finding patterns*

    a. Make an addition table for adding in Alphabitia.
    b. Compare your addition tables with those of others in your group. Is there a "right" size? If you were a teacher, what size addition table would you want your children to make? Justify your choice.
    c. List all the patterns you observe in the table. Listen to other group members describe the patterns that they observed. Add any that didn't appear on your list.

2. *Describing patterns*

    a. Describe one pattern as though you were talking on the phone to a person who was not in class today. Assume that the person has an addition table.
    b. Exchange descriptions with a member of your group. Give feedback to the other person with respect to the accuracy and clarity of that person's statement.

3. *Justifying patterns*   Select one pattern and explain the *why* of that pattern.
4. *Learning Alphabitian addition facts*

    a. Are there any facts that are harder to learn than others? If so, describe what makes them hard.
    b. If you were going to have a quiz on Alphabitian addition facts, what would help you remember them?

5. *Finding missing digits*

    a. This problem has missing digits: B_D + C_ = CBB. Figure out the missing digits alone and show your work.
    b. Compare your solution path and answer with those of other members of your group. Note any new strategies and ideas that you would like to remember.

EXPLORATION 3.3   **Exploring Subtraction in Alphabitia**

1. As you probably discovered in Explorations 3.1 and 3.2, there are different ways in which we can represent and solve addition problems. So too for subtraction problems. Make up two Alphabitian subtraction story problems and solve them. Solve them in a way that makes sense to you—by using manipulatives, diagrams, or numerals. Have at least one of the problems involve regrouping. Focus on making sense of the process, not just on "getting" it.

2. a. What previous knowledge (about numeration systems, place value, addition, and so on) did you need to solve the problem?
   b. If you encountered difficulty, what knowledge or tool(s) got you past the difficulty?
   c. How confident are you that your answer is correct? How might you check your answers other than doing the problem over?

3. Just as you found that there are different kinds of Alphabitian addition problems, there are different kinds of subtraction problems.

   a. List all the subtraction problems made up by the members of your group. Then classify the problems according to type. That is, put problems that feel similar into the same group.
   b. Name and describe each group. The key idea here is for you to look for patterns, similarities, and differences rather than merely "doing it right."

4. Take some time to solve those of the subtraction problems below that your instructor assigns. Solve the problems in a way that *makes sense* to you—by using manipulatives, diagrams, or numerals. If you use manipulatives, draw pictures to record your solution process.

   The goal here is to be able to determine the difference in a way that makes sense to you; there is no one right way to solve these problems. Check each answer before moving on.

   a. DB   b. D0   c. CBD   d. DCB
      −AD   −BB   −ADC   −ABD

   e. C00   f. D00   g. C0B   h. A00
      −BAB   −B0B   −BAD   − CA

   i. B0A   j. B0AC   k. A0B0C l. A0B00
      − BD   − BCD   − B0CA   − D0B

5. Compare your solution path for each problem with those of other members of your group. If some people solved a problem using methods different from yours, can you understand what they did? This is important, because if we honor their intelligence, young children will invent all sorts of interesting ways to solve problems. Many of my students have invented procedures that I have read about in history books.

6. a. Most of you were taught to "carry" and "borrow." That language is discouraged these days. Why do you think this is so? Discuss this matter with your group and write down your group's response. Think of it as a first draft rather than as something that has to be "right."
   b. After listening to other groups' responses, write a second-draft response to the question.
   c. List possible alternative terms to *carrying* and *borrowing*.
   d. After hearing other students' suggestions and rationale, select the words that you would use, and then justify your choices.

EXPLORATION 3.4   **Alternative Algorithms for Addition**

Many of you were surprised to find, in Chapter 2, that the numeration system you grew up with is so recent and that it took so long—thousands of years—to develop. What may surprise you in this chapter is that the algorithms that you were taught to use to add, subtract, multiply, and divide are not the only algorithms in use today and that since the invention of the Hindu-Arabic numeration system, literally hundreds of algorithms have been developed for the four operations.

As I stated in Chapter 2, I think it is important for students to ask and instructors to be able to give legitimate answers to the question "Why do I need to know this?" There are several reasons for you to know not just the standard algorithms but also some alternative algorithms. Before I give my reasons, let me emphasize that "standard" algorithm does not mean the "right" one or even the "best" one, but rather the algorithm that, for various reasons, has become most widespread. First, to the extent that you know not only the hows but also the whys of computing with these four operations, to that extent you will be more helpful to your students when they struggle. A brief example will illustrate my point. My first teaching job was actually teaching high school science. Even though my major in college was mathematics, I had almost the equivalent of a minor in the physical sciences. When I was teaching chemistry and physics, I found that if the students did not understand, most of the time I could only explain it again. The next year, however, when I got a job as a mathematics teacher, I found that if a student was struggling with a mathematics concept, I knew my math "backwards and forwards" and so could explain the concept several different ways. Thus, exploring alternative algorithms will help you to understand these operations and their connection to base 10 much more deeply. Second, I know several elementary teachers who have used some of these alternative algorithms with students who were struggling with standard algorithm, and these students found the alternative algorithms to be helpful. Third, many of my students have told me that they like some of these algorithms better than the ones they were taught. Finally, many students think "it's kinda cool" to see how other people, hundreds of years ago and in different parts of the world today, compute with whole numbers.

The "Scratch" or "Adding Up" Algorithm

At the right, you can see how 564 + 378 is computed with the "scratch" or "adding up" algorithm. You start with the ones place and find the sum of each place. If the sum is less than 10, you simply record the sum at the top. If the sum is 10 or greater, you put the digit from the ones place of the sum above and then cross out the number in the place to the left and increase it by 1. This algorithm is quite probably the origin of the phrase *adding up*.

```
   94
   832
   564
   378
```

1. **a.** Do some more problems on your own until you feel confident using the "adding up" algorithm.
    b. Write directions for using this algoithm and give them to a friend. See whether the friend can add on the basis of your directions. If she or he can, great. If not, have a conversation and find where the directions "went wrong." Keep revising the directions until your directions make sense to your friend. If possible, try the new directions on another friend.

    c. Now that you know how this algorithm works, try to explain the why of each step, as was done for the standard algorithm in the textbook.

The Partial Sums Algorithm

The partial sums algorithm was developed in India over 1000 years ago. It works like this:

```
  3 6 9
  4 7 8
  -----
    1 7
  1 3
  7
  -----
  8 4 7
```

2.   **a.** Do some more problems on your own until you feel confident using the partial sums algorithm.

    b. As before, write directions for using this algorithm and give them to a friend. Have your friend add two numbers using your directions.

    c. As before, justify each step of this algorithm.

The Lattice Algorithm

The lattice algorithm, also popular in the Middle Ages, is one that many of my students like.

    With this algorithm, you can begin at the left or the right. First, you find the sum of the digits in each place; and place each of the sums in the box below, as shown. To determine the answer, you extend the diagonal line segment inside each box and then add the numbers that are in the same "chute," as shown at the right.

3.   **a.** Do some more problems on your own until you feel confident using the lattice algorithm. Try adding three or more numbers.

    b. As before, write directions for using this algorithm and give them to a friend. Have your friend add two numbers using your directions.

    c. As before, justify each step of this algorithm.

A Fourteenth-Century Algorithm

This algorithm is different form the others. Some of my students think this one is bizarre, and some think it is neat. Let us examine the problem 826 + 483 to understand how this algorithm works. Some students have found it helpful to imagine both addends disappearing and the answer appearing in place of the top addend! First, add the digits in the ones place, and place that sum in the ones place of the larger number, replacing the digit that is in that place. That is, replace the 6 in 826 with a 9. Next add the digits in the tens place (2 + 8 = 10). Because our sum is more than 9, we put the 0 in the tens place of the top addend, and we have to increase the hundreds place to the top number by 1. Finally, we add the digits in the hundreds place.

      826   829   909   1309

      483   48   4

Here is another example.

754	762	782	882
128	12	1	

4.   **a.**   Do some more problems on your own until you feel confident using the fourteenth-century algorithm.

   b.   As before, write directions for using this algorithm and give them to a friend. Have your friend add two numbers using your directions.

   c.   As before, justify each step of this algorithm.

EXPLORATION 3.5   **Alternative Algorithms for Subtraction**

As with addition, many different algorithms were developed over the centuries to enable people to do subtraction problems quickly and accurately. We will explore three algorithms here.

The Indian Algorithm

This is one of the earliest algorithms for subtraction and was popular in India almost a thousand years ago. It is called the *reverse method* and has similarities to the standard algorithm in use in the United States in that it involves a "borrowing" step and a "payback" step. Begin at the left and subtract the digits in the furthest left place. You proceed place by place. As long as no regrouping is required, you just sail along, as shown in Step 1. In this example, when we get to the ones place, we encounter the problem of not being able to subtract 4 from 2. As in the algorithm commonly used, we place a 1 above the 2 that is in the ones place, now giving that place a value of 12. Now we do the subtraction (12 − 4) and put the difference of 8 below. Then comes the payback: We cross out the 2 in the tens place of the answer and replace it with a 1. The correct answer is now 418.

Step 1	Step 2	Step 3
632	63^{1}2	63^{1}2
−214	−21 4	−21 4
42	42	42̶8 (1)

1. **a.** Do some more problems on your own until you feel confident using the Indian algorithm.
    b. Write directions for using this algorithm and give them to a friend. See whether your friend can subtract on the basis of your directions. If he or she can, great. If not, have a conversation and find where the directions "went wrong." Keep revising the directions until your directions make sense to your friend. If possible, try the new directions on another friend.
    c. Now that you know how this algorithm works, try to explain the why of each step, as was done for the standard algorithm in the textbook.

The European Algorithm

In 1986, I was helping a friend from New Zealand with some mathematics, and I was stunned to find how he subtracted. Since then, I have discovered that this method is taught in many different countries and that it was taught in many U.S. schools in the earlier part of the twentieth century. In this case, there is no detailed description of how it works, just one example.

984	9 8^{1}4
−368	−3^{7}6̶ 8
	6 1 6

2. **a.** Do some more problems on your own until you feel confident using the European algorithm.
    b. As before, write directions for using this algorithm and give them to a friend. Have your friend subtract two numbers using your directions.
    c. As before, justify each step of this algorithm.

The Treviso Algorithm

The Treviso algorithm appears in what was probably the first arithmetic book published in Europe in 1478 called the *Treviso Arithmetic*. The book contains an explanation of the "new" base 10 system and many different algorithms for adding, subtracting, multiplying, and dividing.

I will give one example, as it appears in the actual manuscript. The problem is $97 - 38$. "(T)ake 8 from 7; this is impossible, but 2, the complement of 8 with respect to 10, added to 7, makes 9, which remainder of this order we write under the 8. Then carry 1 to the 3, making 4, and we have 4 from 9 leaving 5, which write beneath the 2, and the result is 59. Note, also, that this operation of subtraction needs no proof, since we have for its proof the addition already made, that of uniting 38 and 59, making 97."[1]

3. **a.** Do some more problems on your own until you feel confident using the Treviso algorithm.
    b. As before, write directions for using this algorithm and give them to a friend. Have your friend subtract two numbers using your directions.
    c. As before, justify each step of this algorithm.

Looking Back

You have developed your own algorithms for adding and subtracting in Alphabitia, and you have explored other algorithms. Take some time now to reflect on the following questions:

1. What did you learn about addition and subtraction from inventing your own ways of determining the sums or differences or from exploring alternative algorithms?
2. Do you understand the standard algorithms better? If so, what do you understand better now than you did before?
3. Explain how adding and subtracting are connected. Use diagrams and words.
4. When the teacher asked how borrowing and carrying are connected, Uri said, "They are the same process."
    a. What is *the same* about them?
    b. What is *different* about them?

[1] Frank J. Swetz, *Capitalism and Arithmetic: The New Math of the 15th Century* (La Salle, IL: Open Court Publishing Company, 1987), p. 60.

SECTION ◆ **3.2** ◆ **EXPLORING MULTIPLICATION AND DIVISION**

The concepts of and algorithms for multiplication and division are more complex than those for addition and subtraction. Therefore, we will leave Alphabitians explore multiplication and division in base 10. It is crucial that your understanding of multiplication and division be richer than simply knowing how to multiply and how to divide.

EXPLORATION 3.6   **Multiplication from Different Perspectives**

There are many perspectives from which to explore multiplication. We will look at three of them in this exploration.

Skip Counting

Multiplication is often introduced to children as skip counting—that is, counting by 2, by 3, by 4, and so on. A common activity in second grade is to look for patterns in skip counting. This activity is fun and rewarding for college students also.

1. Take out the Patterns in Multiples Charts on pages 67 and 69. In each of the eight charts, color in or lightly shade the squares containing the multiples of the given number. Describe the pattern(s) you see. For example, all multiples of 2 have an even number in the ones place.
2. Now that you have colored in eight different charts, describe any patterns you see among the charts. For example, some patterns appear in all of the charts; some patterns occur in only one or some of the charts. If you have any hypotheses about the reasons for this, write them now.

Multiplication from an Algebraic Perspective

In Step 2 above, we found that some multiples have a vertical pattern, other multiples have diagonal patterns, and still others have both. Let us focus on these patterns from an algebraic perspective as lines and examine the slopes of these lines. (A line's slope is the ratio of its vertical change to its horizontal change.)

3. Determine the slopes of the lines that you see in each table. Let each cell in a table count as 1 unit.
4. What patterns and connections do you see? For example, do certain charts have similar slopes? Why? Can you explain these patterns? For example, why is 1/2, and not 1/3 or 1/4, the slope of one family of lines in the multiples of 8? Write down your explanation.

Multiplication from a Geometric Perspective

Now let us look at multiplication and the multiplication table from a geometric perspective. (A multiplication table is shown in Exploration 3.7.) This exploration comes from an article in *Teaching Children Mathematics*,[2] a journal you will want to subscribe to as soon as possible. Using graph paper, you will make a grid for each cell in the multiplication table up to 10 × 10. For example, 2 × 3 will be ⊞. You may want

---

[2] Barbara E. Armstrong, "Teaching Patterns, Relationships, and Multiplication as Worthwhile Mathematical Tasks," *Teaching Children Mathematics,* 1(7) (1995), pp. 446–450.

to make the entire set of grids yourself, or your instructor may ask each group to construct one master set of grids.

5.   Whether you make the grids alone or in a group, discuss strategies that will enable you to complete the set of grids more efficiently.

6.   Once you have the completed set, "play" with the grids, physically and mentally. Describe any mathematical insights and patterns that you see from your exploration.

Patterns in Multiples Charts for   EXPLORATION 3.6,   Step 1

Multiples of 2

0	1	2	3	4	5	6	7	8	9
10	11	12	13	14	15	16	17	18	19
20	21	22	23	24	25	26	27	28	29
30	31	32	33	34	35	36	37	38	39
40	41	42	43	44	45	46	47	48	49
50	51	52	53	54	55	56	57	58	59
60	61	62	63	64	65	66	67	68	69
70	71	72	73	74	75	76	77	78	79
80	81	82	83	84	85	86	87	88	89
90	91	92	93	94	95	96	97	98	99

Multiples of 3

0	1	2	3	4	5	6	7	8	9
10	11	12	13	14	15	16	17	18	19
20	21	22	23	24	25	26	27	28	29
30	31	32	33	34	35	36	37	38	39
40	41	42	43	44	45	46	47	48	49
50	51	52	53	54	55	56	57	58	59
60	61	62	63	64	65	66	67	68	69
70	71	72	73	74	75	76	77	78	79
80	81	82	83	84	85	86	87	88	89
90	91	92	93	94	95	96	97	98	99

Multiples of 4

0	1	2	3	4	5	6	7	8	9
10	11	12	13	14	15	16	17	18	19
20	21	22	23	24	25	26	27	28	29
30	31	32	33	34	35	36	37	38	39
40	41	42	43	44	45	46	47	48	49
50	51	52	53	54	55	56	57	58	59
60	61	62	63	64	65	66	67	68	69
70	71	72	73	74	75	76	77	78	79
80	81	82	83	84	85	86	87	88	89
90	91	92	93	94	95	96	97	98	99

Multiples of 5

0	1	2	3	4	5	6	7	8	9
10	11	12	13	14	15	16	17	18	19
20	21	22	23	24	25	26	27	28	29
30	31	32	33	34	35	36	37	38	39
40	41	42	43	44	45	46	47	48	49
50	51	52	53	54	55	56	57	58	59
60	61	62	63	64	65	66	67	68	69
70	71	72	73	74	75	76	77	78	79
80	81	82	83	84	85	86	87	88	89
90	91	92	93	94	95	96	97	98	99

Patterns in Multiples Charts for   EXPLORATION 3.6,   Step 1

Multiples of 6

0	1	2	3	4	5	6	7	8	9
10	11	12	13	14	15	16	17	18	19
20	21	22	23	24	25	26	27	28	29
30	31	32	33	34	35	36	37	38	39
40	41	42	43	44	45	46	47	48	49
50	51	52	53	54	55	56	57	58	59
60	61	62	63	64	65	66	67	68	69
70	71	72	73	74	75	76	77	78	79
80	81	82	83	84	85	86	87	88	89
90	91	92	93	94	95	96	97	98	99

Multiples of 7

0	1	2	3	4	5	6	7	8	9
10	11	12	13	14	15	16	17	18	19
20	21	22	23	24	25	26	27	28	29
30	31	32	33	34	35	36	37	38	39
40	41	42	43	44	45	46	47	48	49
50	51	52	53	54	55	56	57	58	59
60	61	62	63	64	65	66	67	68	69
70	71	72	73	74	75	76	77	78	79
80	81	82	83	84	85	86	87	88	89
90	91	92	93	94	95	96	97	98	99

Multiples of 8

0	1	2	3	4	5	6	7	8	9
10	11	12	13	14	15	16	17	18	19
20	21	22	23	24	25	26	27	28	29
30	31	32	33	34	35	36	37	38	39
40	41	42	43	44	45	46	47	48	49
50	51	52	53	54	55	56	57	58	59
60	61	62	63	64	65	66	67	68	69
70	71	72	73	74	75	76	77	78	79
80	81	82	83	84	85	86	87	88	89
90	91	92	93	94	95	96	97	98	99

Multiples of 9

0	1	2	3	4	5	6	7	8	9
10	11	12	13	14	15	16	17	18	19
20	21	22	23	24	25	26	27	28	29
30	31	32	33	34	35	36	37	38	39
40	41	42	43	44	45	46	47	48	49
50	51	52	53	54	55	56	57	58	59
60	61	62	63	64	65	66	67	68	69
70	71	72	73	74	75	76	77	78	79
80	81	82	83	84	85	86	87	88	89
90	91	92	93	94	95	96	97	98	99

EXPLORATION 3.7   **Patterns in the Multiplication Table**

PART 1: Base 10

1.   Examine the multiplication table and describe the patterns that you see.

	1	2	3	4	5	6	7	8	9	10	11	12
1	1	2	3	4	5	6	7	8	9	10	11	12
2	2	4	6	8	10	12	14	16	18	20	22	24
3	3	6	9	12	15	18	21	24	27	30	33	36
4	4	8	12	16	20	24	28	32	36	40	44	48
5	5	10	15	20	25	30	35	40	45	50	55	60
6	6	12	18	24	30	36	42	48	54	60	66	72
7	7	14	21	28	35	42	49	56	63	70	77	84
8	8	16	24	32	40	48	56	64	72	80	88	96
9	9	18	27	36	45	54	63	72	81	90	99	108
10	10	20	30	40	50	60	70	80	90	100	110	120
11	11	22	33	44	55	66	77	88	99	110	121	132
12	12	24	36	48	60	72	84	96	108	120	32	144

2.   Select one pattern that your group will share with the class.

   a. Describe this pattern, as though you were talking on the phone to a friend who missed class today but who has a multiplication table handy. The purpose of your description is simply to get the person to see the pattern.

   b. Exchange descriptions with another group. If they easily interpreted your description, great. If not, revise your description (like a second draft of an essay). Discuss the parts of your description that were not clear to the other group. That is, what was unclear about them, and why do you think the revised wording is better?

   c. What did you learn from describing a pattern and reading another group's description?

3.   Now explain *why* the pattern occurs.

PART 2: Circle clocks

Circle clocks (also called star patterns) have been used by many teachers to introduce many different mathematical concepts and to provide a visual connection for these ideas. The following steps use these circle clocks to give you another perspective on the basic multiplication facts.

   Take out the Base 10 Circle Clocks on page 73. You will draw on one clock for each of the rows in the multiplication table from 2 through 9. The directions are to look only at the units digit. For example, if we look at the multiplication facts in row 2 and write down only the units digit, we have 2, 4, 6, 8, 0, and then the numbers repeat. Start your pencil at 2 on the circle; then draw a line from 2 to 4, then from 4 to 6, and so on.

1. Complete the circle clock for the multiples of 2, and describe the pattern as though you were talking to someone on the phone.
2. Complete the circle clocks for each of the other multiples.
3. What similarities do you see among the different circle clocks? Can you explain why those similarities occur?
4. Predict the shape of the circle clock for the multiples of 11. Explain your reasoning. Then draw the pattern. If your prediction was correct, great. If it wasn't, or if you weren't able to make a prediction, either describe the knowledge that you weren't able to apply or describe what enables you to understand why the pattern is what it is.

PART 3: Other bases

Make a set of multiplication tables in different bases determined by your class or your instructor.

1. a. Describe patterns that seem to be true in all bases.
    b. Describe patterns that are true in some but not all bases.
2. Select one pattern and describe the pattern, as though on the phone to someone who has the tables in front of him or her.
3. Now describe the *why* behind the pattern.

Base 10 Circle Clocks for   EXPLORATION 3.7,   **PART 2**

Base 10 Circle Clocks

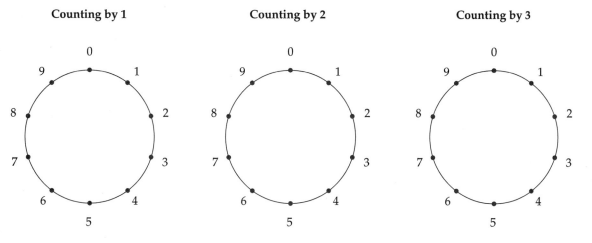

Counting by 1

Counting by 2

Counting by 3

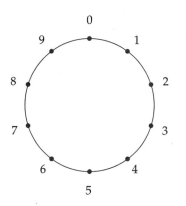

Counting by 4

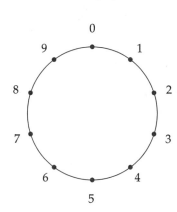

Counting by 5

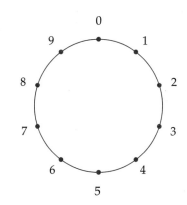

Counting by 6

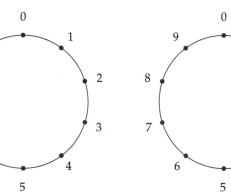

Counting by 7

Counting by 8

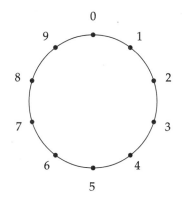

Counting by 9

EXPLORATION 3.8   **Understanding the Standard Multiplication Algorithm**

Using your understanding of our base 10 numeration system and your understanding that multiplication can be seen both as repeated addition and as a rectangular array, you can now understand *why* our modern multiplication algorithm works. We will begin by examining simple multiplication problems and then gradually work up to bigger ones.

1.   What if you forgot what 8×7 was? Describe two different ways in which you could figure it out that would be quicker than adding 7 eight times or adding 8 seven times. After the class discussion, describe other ways that you like.

2.   For each of the problems below, assume that you know multiplication facts only up to 10×10.

   a.   Describe how you could determine 13×7. After the class discussion, describe other ways that you like.

   b.   Describe how you could determine 17×12. After the class discussion, describe other ways that you like.

   c.   Describe how you could determine 19×13. After the class discussion, describe other ways that you like.

   d.   Describe how you could determine 34×23. After the class discussion, describe other ways that you like.

3.   Using a copy of the base 10 graph paper at the end of the book, represent 34×23 as a rectangle. Determine the answer using the idea of covering this amount with base 10 blocks — first flats, then longs, then units. Look at how you determined the product, and then look at how you determine the product using the standard algorithm (the usual multiplication method). Now describe any new insights you have about why the algorithm works.

4.   Using the problem below, explain why each step in the standard multiplication algorithm works.

$$\begin{array}{r} 56 \\ \times 37 \\ \hline 392 \\ 168 \\ \hline 2072 \end{array}$$

   a.   We begin by multiplying 7 and 6. How do we know to multiply these two digits and not, for example, 5 and 6?

   b.   Then, instead of placing the product (42) below, we place only the 2 below. Why is this?

   c.   Next, we find the product of 5 and 7. Why are we doing this step next?

   d.   Then we add the 4 from the previous step to the 35 we get from 5 times 7, and we place the 39 to the left of the original 2. What does this 392 represent?

   e.   Next we multiply 3 and 6. How do we know to multiply these two digits?

   f.   Why don't we place the 8 from this multiplication under the 2 instead of under the 9? That is, why do we "move over"?

   g.   Finally, we add 392 and 168. If this is multiplication, then why are we adding? What do those numbers represent?

EXPLORATION 3.9 **Alternative Algorithms for Multiplication**

"Before the introduction of the Arabic notation, multiplication was difficult, and the division even of integers called into play the highest mathematical faculties. Probably nothing in the modern world could have more astonished a Greek mathematician than to hear that, under the influence of compulsory education, the whole population of Western Europe . . . could perform the operation of division for the largest numbers. This fact would have seemed . . . a sheer impossibility."[3]

Egyptian Duplation

How do you think people multiplied before base 10? Imagine finding 35×28 with the Egyptian numerals we saw in Section 2.3 of the text—that is, ∩∩∩||||| times ∩∩||||||||. The Egyptians, who did not have a numeration system with a base, developed an algorithm for multiplying based on the idea of multiplication as repeated addition and using doubling. This was the most practical way to multiply in the Western world before base 10 was adopted.

The Egyptians would have done the following to multiply 35×28:

∩∩∩|||||

∩∩∩∩∩∩

𝓞∩∩∩∩

𝓞𝓞∩∩∩∩∩∩∩

𝓞𝓞𝓞𝓞𝓞∩∩∩∩∩

You might be able to figure out the Egyptian method more easily if you use Hindu-Arabic numerals.

35	
70	
140	
280	
560	

The Egyptians would then have added

𝓞∩∩∩∩

𝓞𝓞∩∩∩∩∩∩∩

𝓞𝓞𝓞𝓞𝓞∩∩∩∩∩

To get 35×28, add

$$
\begin{array}{r}
140 \\
280 \\
+560 \\
\hline
980
\end{array}
$$

to come up with

𝓞𝓞𝓞𝓞𝓞𝓞𝓞𝓞𝓞∩∩∩∩∩∩∩∩

1. **a.** Do some more problems on your own until you feel confident using the Egyptian duplation algorithm.
   b. Write directions for using this algorithm and give them to a friend. See whether the friend can get the correct answer on the basis of your directions. If she or he can, great. If not, have a conversation and find where the directions "went wrong." Keep revising the directions until your directions make sense to your friend. If possible, try the new directions on another friend.
   c. Now that you know how this algorithm works, try to explain the why of each step, as was done for the standard algorithms in the textbook.

---

[3]Alfred North Whitehead, *Introduction to Mathematics*, New York, 1911, p. 59, cited in Robert Moritz, *On Mathematics* (New York: Dover Publications, 1914), p. 198.

The Lattice Algorithm

Just as there was a lattice algorithm for addition, there is also one for multiplication. See whether you can figure out how it works for 45×28. Just as there are four partial products in the standard algorithm, so too in this algorithm. One of my grandmothers told me that this is how she learned multiplication when she was a little girl.

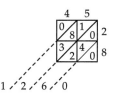

2. **a.** Do some more problems on your own until you feel confident using this algorithm. Note that this and the next two algorithms can be applied to larger problems, such as multiplying 2-digit by 3-digit numbers or 3-digit by 3-digit numbers.
    b. As before, write directions for using this algorithm and give them to a friend. Have your friend multiply two numbers using your directions.
    c. As before, justify each step of this algorithm.

The "Cross Product" or "Lightning" Algorithm

This algorithm first appears in the Treviso Arithmetic. It became one of the more popular algorithms for multiplying. The direct translation of the Treviso instructions for multiplying 56 by 48 is given below.

   "(M)ultiply 6 times 8, making 48; write 8 under the units and reserve 4. Then multiply crosswise, thus: 4 times 6 are 24, and 5 times 8 are 40; add 24 and 40, giving 64, and add the 4 which was carried making 68; write 8 and carry 6. Now multiply the tens by the tens, thus: 4 times 5 are 20, and 6 to carry are 26, which is written in its proper place. The result is therefore 2688. We then say that 48 times 56 are 2688. In this same way you can perform all other cross multiplications."[4]

$$
\begin{array}{r}
56 \\
\times 48 \\
\hline
2688
\end{array}
$$

3. **a.** Do some more problems on your own until you feel confident using this algorithm.
    b. As before, write directions for using this algorithm and give them to a friend. Have your friend multiply two numbers using your directions.
    c. As before, justify each step of this algorithm. *Hint*: Express the numbers in expanded form, such as 56 as 50 + 6.
    d. Can you adapt this algorithm for 3-digit by 2-digit problems? For 2-digit by 3-digit problems?

---

[4] *Ibid*, p. 75.

The "Door Junction" or "Sliding" Algorithm

A whole family of methods that were very popular in India go by these names. The name comes from the action of sliding the numerals in the multiplier to the left after each step. Many students report finding this algorithm to be more difficult to understand than the others. If you find this to be true, then read the hint at the end of the example following. I strongly encourage you to do this problem with pencil-and-paper as you read the steps.

1 5 ↗↗ 1 3 4	Step 0: Setting up the problem. Place the multiplier so that the ones place of the multiplier is diagonally to the right above the ones place of the multiplicand.
1 5 1 3 6 0	Step 1: Do the two partial products: 4 × 5 and 4 × 1. 4 × 5 = 20. Place the 0 below the 5, and carry the 2. 4 × 1 = 4 which we add to the 2 that we carried, giving us 6. Replace the 4 in what is now the 10s place by the 6.
1 5 ↗↗ 1 3 6 0 1 5 1 5 1 0	Step 2: Slide the 15 one place to the left. Do the next two partial products: 3 × 5 and 3 × 1. 3 × 5 = 15 which we add to the 6, giving us 21. Replace the 6 in the 10s place by the 1, and carry the 2. 3 × 1 = 3 which we add to the 2 that we carried, giving us 5. Replace the 3 in the 100s place by the 5.
1 5 ↗↗ 1 5 1 0 1 5 2 0 1 0	Step 3: Slide the 15 one place to the left. Do the next two partial products: 1 × 5 and 1 × 1. 1 × 5 = 5 which we add to the 5, giving us 10. Replace the 5 in the 100s place by the 0, and carry the 1. 1 × 1 = 1 which we add to the 1 that we carried, giving us 2 in the 1000s place.

Hint: If you do this problem with the traditional algorithm you will perform the same six partial products and in the same order—first, 4 × 5 and 4 × 1; second, 3 × 5 and 3 × 1; third, 1 × 5 and 1 × 1.

With this algorithm, we slide the digits in the multiplier 1 place to the left to help us "keep place."

1 5	1 5	1 5
1 3 4	1 3 4	1 3 4

Let us examine another example: 456×23

2 3 456	Step 0: Setting up the problem. Place the multiplier so that the ones place of the multiplier is diagonally to the right above the ones place of the multiplicand.

  2 3 4 5 3 8	Step 1: Do the two partial products: 6×3 and 6×2. $6 \times 3 = 18$. Place the 8 below the 3, and carry the 1. $6 \times 2 = 12$ which we add to the 1 that we carried, giving us 13. Replace the 6 in the 10s place by the 3, and carry the 1 to the second part of the next step.

2 3 4 5 3 8 2 3 4 2 8 8	Step 2: Slide the 23 one place to the left. Do the next two partial products: 5×3 and 5×2. $5 \times 3 = 15$ which we add to the 3, giving us 18. Replace the 3 in the 10s place by the 8, and carry the 1. $5 \times 2 = 10$ which we add to the 1 that we carried, giving us 11, which we add to the 1 that we carried from the previous step, giving us 12. Replace the 5 in the 100s place by the 2, and carry the 1 to the second part of the next step.

2 3 4 2 8 8 2 3 10 4 8 8	Step 3: Slide the 23 one place to the left. Do the next two partial products: 4×3 and 4×2. $4 \times 3 = 12$ which we add to the 2, giving us 14. Replace the 2 in the 100s place by the 4, and carry the 1. $4 \times 2 = 8$ which we add to the 1 that we carried, giving us 9, which we add to the 1 that we carried from the previous step, giving us 10. Replace the 4 in the 1000s place by the 10, and we are done.

4.   **a.**   Do some more problems on your own until you feel confident using the sliding algorithm.

   b.   As before, write directions for using this algorithm and give them to a friend. (In this case, pick a *very good friend*!) Have your friend multiply two numbers using your directions.

   c.   As before, justify each step of this algorithm.

EXPLORATION 3.10   **Connecting Multiplication and Number Sense**

We have spent some time developing an understanding of why the standard multiplication algorithm works. Another related need is to begin to connect the algorithms to the idea of place value, which, as we discovered in Chapter 2, is an essential characteristic of our Hindu-Arabic numeration system. This exploration is powerful for many students, for it helps them develop some of the number sense they need when solving real-world problems.

Where Do the Digits Go to Make the Biggest Product?

This exploration is to be done with a partner. The object is to place the digits 1, 2, 3, 4, and 5 in the boxes in such a way as to obtain the greatest product.

$$\begin{array}{r} \square\ \square\ \square \\ \times\quad \square\ \square \\ \hline \end{array}$$

In this exploration, we will be using guess–check–revise.

1.   With your partner, take a few moments to decide upon your first guess. Briefly write your reasoning for your guess. Stop and think what digit you want in the hundreds place and whether it matters which digits go in the two tens places and which digits go in the two ones places.

2.   Determine the product longhand; do not use a calculator. Then stop and reflect on this guess. Did you learn anything from the guess?

3.   Now take a few moments to decide upon your second guess. Write your second guess, and briefly explain your reasoning for this guess.

4.   Continue this process until you feel that you have found the combination that produces the greatest product.

5.   Merge with another pair. If you have the same combination, share with each other (and then write) strategies that helped and what you have learned. If one pair has a greater product, have that pair share the reasoning that led to that hypothesis.

6.   Merge with another group of four or join the whole class, according to your instructor's directions, to discuss the problem. Write what you learned from the discussion.

EXPLORATION 3.11   **Different Models of Division**

1. Make up a story problem for 15 ÷ 3. Assume that you are a young child who does not yet know division. Describe how you would determine the answer. You may use pictures to describe your solution path, but you are not required to.

2. Compare your stories with those of other members of your group. Just as we found that there are different models for addition, subtraction, and multiplication, there are different models for division.

    a. Sort your stories into two or more groups. Explain the common characteristic(s) of the stories in each group.
    b. Give a name to each of the groups and briefly justify that name.
    c. After listening to the class discussion, name and briefly define each of the models for division.

3. Consider the problem 92 ÷ 4. Make up a story problem for these numbers that fits each of your models. Still assuming that you don't know the formal idea of division, describe how you would determine the answer.

EXPLORATION 3.12   **Understanding the Standard Algorithm**

Consider the following story problem: There are three fifth-grade classrooms. The janitor has taken all of the desks out of the classrooms, has waxed the floors, and is now ready to put the desks back in the classrooms. The only problem is that he has forgotten how many go into each classroom. If there are 72 desks, how many go into each classroom?

1. When children who don't know the pencil-and-paper procedure are asked to solve this problem with base 10 blocks, they often do so in the seven steps shown in the middle column of the table below. The right column asks you to describe what is happening mathematically for each step. Note that some students and teachers might combine some of the steps below into fewer steps; for example, they might see my Steps 1 and 2 as a single step. Note that some students will arrive at 2 longs in Step 1 by using the missing factor model of division; some students will arrive at 2 longs in each classroom from the partitioning model and others by the repeated subtraction model. This diversity of solution paths makes teaching both exciting and difficult!

Step	What is going on at the physical level	What is going on mathematically
1	We determine that we can put 2 longs in each classroom.	
2	We physically place 2 longs in each classroom. (Often the teacher will have three pieces of construction paper, one to represent each classroom.)	
3	We look at what is left: 1 long and 2 singles.	
4	We exchange the 1 long for 10 singles.	
5	We now have 12 singles.	
6	We determine that we can put 4 singles in each classroom.	
7	We now know that the answer is 2 longs and 4 singles in each classroom—that is, 24 chairs in each classroom.	

2. Most young children are able to do this problem at the physical level. Mathematical power comes from being able to connect the physical actions to the pencil-and-paper algorithm—that is, to get at the *whys* behind the *whats*. The middle column in the table below describes the verbalization of each step in the pencil-and-paper algorithm. The right column asks you to explain the why for each step. Note that there is not a one-to-one correspondence between the seven steps in the physical solution of the problem above and the six steps in the pencil-and-paper solution following.

$\begin{array}{r} 24 \\ 3\overline{)72} \\ \underline{6} \\ 12 \\ \underline{12} \end{array}$	**What we do**	**Why we do it**
	a. 3 "gazinta" 7 two times	a. What is going on mathematically when we say "gazinta"?
	b. Place the 2 above the 7.	b. Why do we place the 2 above the 7?
	c. $3 \times 2 = 6$. Place the 6 below the 7.	c. Why do we multiply 3×2?
	d. Subtract 6 from 7.	d. Why do we subtract?
	e. Bring down the 2.	e. What is going on mathematically when we "bring down" the 2?
	f. 3 "gazinta" 12 four times. Place the 4 above the 2, Next, $3 \times 4 = 12$. Place the 12 below the other 12.	f. Why do we repeat these three steps again: gazinta, multiply, subtract?

Next, we will examine two larger problems that expose two difficulties students commonly struggle with.

3. A company has received 252 computers and will ship the same number to each of its 4 stores. How many computers does each store get? The middle column of the table below contains selected steps in the algorithm. The right column asks you to justify the step.

$\begin{array}{r} 63 \\ 4\overline{)252} \\ \underline{24} \\ 12 \\ \underline{12} \end{array}$	**Algorithm**	**Justification**
	a. 4 doesn't go into 2.	a. What is going on mathematically here?
	b. 4 goes into 25 six times.	b. What is going on mathematically here?
	c. $4 \times 6 = 24$; put the 6 above the 5.	c. How do you know where to put the 6?
	d. Subtract 24 from 25.	d. Why are we subtracting?
	e. Bring down the 2.	e. What does this mean mathematically?

4. Many students have trouble when there are zeros in the quotient. Explain these steps in the long-division problem below.

$\begin{array}{r} 304 \\ 27\overline{)8208} \\ \underline{81} \\ 108 \\ \underline{108} \end{array}$	**Algorithm**	**Justification**
	a. $27 \times 3 = 81$; put the 3 above the 2.	a. How do you know where to put the 3?
	b. Subtract $82 - 81$ and bring down the 0.	b. Why do we bring down the 0, since we are bringing down nothing?
	c. Since 27 doesn't go into 10, put a 0 above the 0.	c. Why?
	d. Bring down the 8.	d. Why?

EXPLORATION 3.13   **Alternative Algorithms for Division**

Before base 10, division was a very difficult and time-consuming process. Most of the pre-base 10 division algorithms involved some aspect of missing factor. For example, if we are dividing 756 by 23, we ask what number times 23 will give us 756. Some of the algorithms were like the Egyptian duplation in reverse. Even with the development of base 10, it was many centuries before the most commonly used algorithm was invented. As you may have already discovered, the standard division "gazinta" algorithm is not intuitively obvious. That is, it is not a procedure that was easily developed. After all, there are many operations within this algorithm—dividing, muliplying, subtracting, and "bringing down."

The Scaffolding Algorithm

Many students have trouble with long division. Common problems are poor multiplication facts, not being able to remember the sequence (divide, multiply, subtract, bring down), and misplacing the digits in the quotient. The standard long-division algorithm can be unforgiving, as the example below shows. After repeated failure with the standard algorithm, many students simply give up; many of my students in this course have told me of bad experiences learning the division algorithm.

$$
\begin{array}{r}
69 \text{ R } 24 \\
8\overline{)576} \\
48 \\
\hline
96 \\
72 \\
\hline
24
\end{array}
$$

An alternative algorithm, called the scaffolding algorithm, has been helpful in giving students a sense of success, which builds their confidence. It also develops their ability to use guess–check–revise and reinforces their multiplication skills.

Here is how it might work with the problem $576 \div 8$, which might represent 576 students grouped into 8 teams or 576 oranges divided into bags with 8 oranges per bag. In the former case, the student is asked, "About how many students do you think would be on each team?" In the second case, the student is asked, "About how many bags do you think you can make?"

Let's say the student's first guess is 60—that is, 60 students per team or 60 bags of oranges. The student multiplies 60×8 to see how many students or oranges have been "used up" and how many are left to be dealt with. The 60 is placed at the top. (Note: the diagram for each step is on the next page.)

The student is now asked the same question with the remaining 96. Let's say the student guesses 10. The 10 is placed above the 60 (that is, in the "answer" space), and the student takes away the 10 groups of 8, or 80.

We now have 16 (students or oranges) to deal with. The student sees that $8 \times 2 = 16$ and places the 2 in the answer space. The student now adds $60 + 10 + 2$ to get the answer.

At the far right is another example of this algorithm; note that the actual numbers in the answer space and the number of steps depend on the student's guess.

```
                              1
                    2         6
          10       10        20
  60      60       60        50
8)576   8)576    8)576    64)4953
  480     480      480       3200
   96      96       96       1753
           80       80       1280
           16       16        473
           16       16        384
            0        0         89
                               64
                               25
```

1. **a.** Do some more problems on your own until you feel confident using the scaffolding algorithm.
    b. Write directions for using this algorithm and give them to a friend. See whether the friend can add on the basis of your directions. If he or she can, great. If not, have a conversation and find where the directions "went wrong." Keep revising the directions until your directions make sense to your friend. If possible, try the new directions on another friend.
    c. Now that you know how this algorithm works, try to explain the why of each step, as was done for the standard algorithm in the textbook.
    d. What advantages might this algorithm offer in teaching?

The Galley or Scratch Algorithm

A method of long division called the galley method was popular in Europe in the Middle Ages. It is a modification of an algorithm that traces back to eighth-century India. It was called the galley method because the scratches caused the diagram to resemble a sail, and a galley is a kind of ship. The example following shows how this method would be used to divide 1750 by 14.

Computation	Explanation
1750 14	The problem begins by dividing 17 by 14.
3 1750 1 14 14	The quotient of 1 is placed at the right. The difference is placed above the 7. Then the 17 and the 14 are crossed out and the divisor (14) is moved one place to the right but is written diagonally (probably to save space—paper was very expensive in those days!)
3 1750 12 14 14	Then 35 is divided by 14, giving a quotient of 2, which is placed to the right of the 1.
37 1750 12 14 14 14	The 35 is replaced by the remainder of 7. The 14 is crossed out and moved over 1 place.
37 1750 125 14 14 14	Then 70 is divided by 14, giving a quotient of 5.

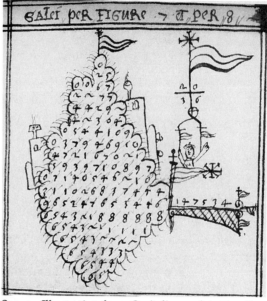

Source: Illustration from *Capitalism and Arithmetic*
by Frank Swetz. Open Court Publishing Co.,
LaSalle, IL.

2.   **a.** Do some more problems on your own until you feel confident using the
galley algorithm.

      b. As before, write directions for using this algorithm and give them to a friend.
Have your friend divide two numbers using your directions.

      c. As before, justify each step of this algorithm.

EXPLORATION 3.14   **Dealing with Remainders**

Real-life problems involving division generally require not just the ability to do the computation but also the ability to interpret the result. This exploration enables you to explore the issue of remainders more deeply.

1.   The Allentown Elementary School is going to the Science Museum in a nearby city. There are 369 students going, and each school bus can hold 24 students. How many school buses will be needed? Do this problem by yourself and then discuss your answer with members of your group.

2.   Make up a story problem involving 31 divided by 4 in which the answer is

    a. 7   **b.** 8   **c.** $7\frac{3}{4}$ **d.** 7 remainder 3

3.   After listening to other students' story problems, what did you learn about division from this exploration?

Looking Back

As you look back on the explorations with addition, subtraction, multiplication, and division, stop and reflect on what you have learned.

1.   Briefly describe at least three important learnings. Below are four questions to start the reflection process.

    ■ Did you learn more about what one or more of the operations means?

    ■ Do you better understand how one or more of the algorithms work?

    ■ Do you better understand how two or more of the operations are connected?

    ■ Did working through these explorations cause you to understand place value more deeply or to understand how zero works in base 10?

2.   Select and discuss one of these learnings. Your discussion needs to contain these three aspects:

    a. A description of *what* you learned

    b. Within this description, an explanation of the "why" behind the "what"

    c. An explanation of why this learning seems important to you

EXPLORATION 3.15   **Developing Operation Sense: String Art**

Many real-life problems are what NCTM calls "nonroutine, multistep problems"—that is, problems that require more than one operation. In many of these situations, it is not obvious which operation to use. Unfortunately, many students have learned "tricks" that help them decide what operation to use—for example, "*of* means multiply;" "if you see the word *more,* then subtract the numbers," and so on. This exploration contains problems that require various operations and it requires you to apply your understanding of the operations in a nonroutine way: They require thinking as opposed to remembering tricks.

An elementary teacher, we'll call him Stanley, has found out about string art and has made a string art sculpture like the one below.

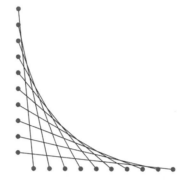

1. **a.** What do you see in the design?
    b. Make a copy of the design. Share your strategy for making the design with other members of your group. Did some members find ways to make the drawing that were easier than the ways others found?
    c. Having made the drawing yourself, do you see anything in the design that you didn't see before?
    d. Make another design of our own. *Hint:* Place a blank sheet of paper over the graph paper at the end of this book in order to place the "nails" nicely.

2. Suppose Stanley wants his whole class to make these sculptures.

    a. How many nails will he need for a class of 24 students?
    b. If the nails are 1 inch apart, approximately how much string will he need?
    c. Stanley wants to use four different colors of tapestry thread, because there are four pairs of strings that have the same length. How much of each color will he need for a class of 24? How much will the string cost?
    d. A friend of Stanley's has an extra piece of plywood; it is 6 feet long and $3\frac{1}{2}$ feet wide. Stanley wants each smaller piece of wood to be big enough so that the nails are 1 inch from the edge of the plywood. Will this piece of wood be big enough to make 24 smaller pieces?
    e. For a similar project, Stanley needs 684 nails, which come in packages of 18. How many packages will he need?

3. Stanley has decided to have his students make a new string sculpture. He has determined that each group will need 75 inches of string. He has a roll of string that is 912 inches, and he will have 12 pairs of students. Does he have enough string, or will he need to buy more?

4. Patterns in string art sculptures.

    a. How many intersection points are there in an 8 by 8 string sculpture?
    b. How many intersection points are there in a 9 by 9 string sculpture?
    c. How many intersection points are there in an *n* by *n* string sculpture?

EXPLORATION 3.16   **Operation Sense in Games**

GAME 1: Greatest amount

Materials
- One die with the following numbers: 0, 1, 2, 3, 4, 5. (You can simply tape a 0 over the 6 on a standard die.)

Directions for playing the game
- Roll the die four times and record the numbers.
- The object of the game is to place one number in each blank so that the value of the given expression is as great as possible.

1.   The object is to make the greatest amount in the following expression.

$$\square \times \square + \square - \square$$

   a.   Before the first game, stop to think a moment about strategy. Can you describe a master strategy that you think will work for all cases? For example, where will you place the largest number? What if you roll a zero? Write down your master strategy for placing the numbers to make the greatest amount.
   b.   Play the first round in the following manner. Roll the die four times and record the numbers. Take a moment to think by yourself where you will place the numbers. Go around the group and have each member present a choice and the reasoning behind it. Once you have determined the largest possible value for these four numbers, roll the die four times again and repeat the process. Play as many rounds as you need until each member has found a strategy that that member believes makes sense and will produce the greatest possible answer each time.
   c.   Write down your master strategy if it has changed since the beginning of the game.

2.   The object is to make the greatest amount in the following expression.

$$\square \times \square \div \square + \square$$

   a.   Before rolling the die, use your reasoning abilities: Do you think the same master strategy you developed in Step 1 will apply, or do you need to modify it? Write down your thoughts.
   b.   Play several rounds, recording your numbers and your reasoning.
   c.   Write down your master strategy if it has changed since the beginning of the game.

3.   Again, the object is to make the greatest amount in the following expression.

$$\frac{(\square \times \square)}{(\square - \square)}$$

   a.   Before rolling the die, use your reasoning abilities: Write your first draft of a master strategy for this expression.
   b.   Play several rounds, recording your numbers and your reasoning.
   c.   Write down your master strategy if it has changed since the beginning of the game.

GAME 2: Smallest amount

Game 2 is played in the same manner as Game 1 except that now the object of the game is to make the value of the given expression as *small* as possible.

1.   The object is to make the smallest amount in the following expression.

$$\Box \times \Box + \Box - \Box$$

    a.   Before rolling the die, use your reasoning abilities: Write your first draft of a master strategy for this expression.
    b.   Play several rounds, recording your numbers and your reasoning.
    c.   Write down your master strategy if it has changed since the beginning of the game.

2.   The object is to make the smallest amount in the following expression.

$$\Box \times \Box \div \Box + \Box$$

    a.   Before rolling the die, use your reasoning abilities: Write your first draft of a master strategy for this expression.
    b.   Play several rounds, recording your numbers and your reasoning.
    c.   Write down your master strategy if it has changed since the beginning of the game.

3.   Again, the object is to make the smallest amount in the following expression.

$$\frac{(\Box \times \Box)}{(\Box - \Box)}$$

    a.   Before rolling the die, use your reasoning abilities: Write your first draft of a master strategy for this expression.
    b.   Play several rounds, recording your numbers and your reasoning.
    c.   Write down your master strategy if it has changed since the beginning of the game.

GAME 3: Target

Here is a game that is adapted from the *Fifth-Grade Book* in the NCTM Addenda Series, Grades K–6 (p. 48).

Materials
- Three dice:
  1, 2, 3, 4, 5, 6
  1, 2, 3, 4, 5, 6
  7, 8, 9, 10, 11, 12

Directions for playing the game
- Pick a target number.
- Roll the three dice.
- Use the three numbers you roll and any combination of operations to get as close as possible to the target number you have chosen.

1.   Let's say the target number was 27 and you rolled 3, 5, 10.

    a.   Try on your own to get as close to 27 as you can.
    b.   What strategies did you use? Did your understanding of the four operations help you? How?

    c. After you write your responses, discuss your responses with other members of your group.

2.   With your partners, select a target number and roll the dice.

    a. Work by yourself for a few minutes to determine your response.

    b. Compare answers and strategies with other members of your group. If one or more members are closest to the target, discuss strategies. Note any new strategies or methods that you prefer.

    c. Play several more rounds.

Looking Back on Exploration 3.16

In what ways did playing these games help you to learn about how the operations are related?

SECTION **3.3**   **EXPLORING MENTAL ARITHMETIC AND ESTIMATION**

Number sense is like common sense with respect to numbers. There is much evidence that the average American does not have much number sense and that this should concern us. The following explorations present you with opportunities to develop different aspects of number sense.

EXPLORATION 3.17   **How Many . . .? How Far . . .?**

1. Write down your guess for each of the following numbers. In each case, if there was any reasoning involved (that is, other than a number simply popping into your head), briefly describe the reasoning.

    a. What is the population of the city or town in which your college is located?
    b. What is the population of the United States?
    c. What is the population of the world?
    d. What is the number of K–12 schoolchildren in the United States?
    e. What is the number of K–12 schoolteachers in the United States?
    f. What is the number of doctors in the United States?
    g. Decide upon a relatively short distance that all members of the class can relate to—for example, the distance from one spot on campus to another. Decide upon a unit of measurement: feet, miles, minutes. Then estimate the distance.
    h. How far is it from New York to Los Angeles?
    i. How far is the Earth from the sun?

2. Discuss how you might find answers for each of the questions in Step 1.
3. Under what circumstances does it matter whether some (even many) students' guesses are "way off"?
4. Rough guesses.

    a. Do you think it is important that people have a rough idea of certain amounts? If you believe it is, explain why. If you believe it is not, explain why not.
    b. How rough is rough? For example, what guesses for the population of the United States would be "reasonable"? How did you determine what is reasonable?

EXPLORATION 3.18   **Mental Arithmetic**

Many of the numbers we see in daily life are estimates, and much of the arithmetic we do involves estimation. There are several skills needed for good estimation, one of which is being able to do mental arithmetic—being able to do simple computations in your head rather than doing everything with pencil and paper or with a calculator. The following explorations are designed to develop your mental arithmetic abilities. Before you begin, take out the four tables that appear on pages 95–98.

PART 1: Mental addition

1. Use the Mental Addition table on page 95. Do in your head the eight computations shown in the table. Briefly note the strategies that you used, and try to give names to them.
    One mental tool all students have is being able to visualize the standard algorithm in their heads. For example, for part (a), you could say: "9 + 7 = 16, carry the one, then 5 + 3 = 8 + the carried 1 makes 9; the answer is 96." However, because you already own that method, I ask you not to use it here but to try others instead.
2. Share your strategies in your small group. Note any strategies that you heard that you did not use but would like to use.
3. In your group, select three or four strategies to describe to the class. Make up a name for each strategy.
4. After hearing the class presentations, write down the strategies that you like best.

PART 2: Mental subtraction

Most people find it more difficult to determine exact answers to subtraction problems in their heads. However, if you think about the various models for subtraction that we have discussed, there are many possibilities: take-away, comparison, adding up, and the fact that the difference between two numbers tells us how far apart they are on the number line, to name but a few.

1. Use the Mental Subtraction table on page 96. Do in your head the six computations shown in the table. Briefly note the strategies that you used other than simply doing the standard algorithm in your head. (Once again, this strategy is not bad; it's just one that everyone already has.) Try to give names to your strategies.
2. Share your strategies in your small group. Note any strategies that you heard that you did not use but would like to use.
3. In your group, select three or four strategies to describe to the class. Make up a name for each strategy.
4. After hearing the class presentations, write down the strategies that you like best.

PART 3: Mental multiplication

1. Use the Mental Multiplication table on page 97. Do in your head the six computations shown in the table. Briefly note the strategies that you used other than simply doing the standard algorithm in your head. Try to give a name to each strategy.
2. Share your strategies in your small group. Note any strategies that you heard that you did not use but would like to use.
3. In your group, select three or four strategies to describe to the class. Make up a name for each strategy.

4.   After hearing the class presentations, write down the strategies that you like best.

PART 4: Mental division

1.   Use the Mental Division table on page 98. Do in your head the five computations shown in the table. Briefly note the strategies that you used, and try to give names to them.
2.   Share your strategies in your small group. Note any strategies that you heard that you did not use but would like to use.
3.   In your group, select three or four strategies to describe to the class. Make up a name for each strategy.
4.   After hearing the class presentations, write down the strategies that you like best.

Looking Back on Exploration 3.18

Are there any strategies that work for more than one operation? Are there any strategies that are similar?

Mental Addition Table for EXPLORATION 3.18, PART 1

Sum	What you did	Name
a. 39 +57		
b. 27 +58		
c. 78 +25		
d. 46 +19		
e. 625 +147		
f. 588 +225		
g. 790 +234		
h. 8734 + 467		

Mental Subtraction Table for EXPLORATION 3.18, PART 2

Difference	What you did	Name
a. 65 −28		
b. 71 − 39		
c. 80 −36		
d. 324 −275		
e. 152 − 87		
f. 1000 − 378		

Mental Multiplication Table for   EXPLORATION 3.18,   PART 3

Product	What you did	Name
a.　82 　　× 5		
b.　24 　　×25		
c.　23 　　×12		
d.　638 　　× 2		
e.　70 　　×40		
f.　83 　　×24		

Mental Division Table for EXPLORATION 3.18, PART 4

Quotient	What you did	Name
a. $40\overline{)8000}$		
b. $80\overline{)40,000}$		
c. $72\overline{)3600}$		
d. $7\overline{)1463}$		
e. $6\overline{)174}$		

EXPLORATION 3.19   **Having a Party**

Let's say you are having a party and expect 65 people. You want to serve juice. How many bottles of juice should you buy? What additional information do you need to have in order to answer this question? Also, what assumptions or factors will guide your decision? For example, I would assume that the guests will drink more if I serve the juice in large glasses than if I serve it in small glasses. Work on similar preliminary details in Steps 1 and 2 without a calculator.

1. What assumption or factors do you need to agree on before you do any computation?
2. Determine the data that you will need in order to make your estimate. For example, the juice comes in 32-ounce bottles, or 64-ounce bottles, or 1-liter bottles.
3. Now do the actual computation, explaining the steps in your solution.
4. Listen to other groups' presentations.

EXPLORATION 3.20   **How Much Did They Save?**

Several years ago, the Education Department at Keene State College determined that we could save a substantial amount of money if we purchased a risograph machine (a high-tech mimeograph machine) and used it for duplicating instead of the copying machine. When we use the copying machine, our department is charged 5¢ per copy; with the risograph machine, we are charged 2¢ per copy.

In order to budget better for the future, the department also wanted to get a sense of how much the machine was used, by whom it was used, and how large the runs were. For example, when I duplicate a test for my math class, I make 30 copies. However, when I duplicate an observation form for the 12 sections of the Learning Theory class, I run off 250 copies. The risograph machine has a counter, and professors were asked to record the beginning number, the ending number, the number of copies, the date, and their name. One line in that sheet would look like this:

Date	Begin	End	Number	Professor
9/23	34,567	34,867	300	Bassarear

I took the following data from one week:

- Counter at the beginning of the week: 127,322
- Counter at the end of the week: 140,001
- Total number of runs: 58

I asked myself the four questions, which I would like you to address in the following manner.

1. Take out the table on page 101. First, determine your estimate for each question in the table. (*Note:* An estimate means a number that you can get mentally within a few seconds.) Briefly note what you did in your head. Then go around the group and have each person describe how she or he obtained an estimate. Describe any strategies that you did not use but that you like. .

2. Describe any assumptions you are making when you do this problem.

Table for EXPLORATION 3.20, How Much Did They Save?

Question	Estimate	Brief description of how you got your estimate
a. What was the total number of copies run that week?		
b. What was the average number of copies per day?		
c. What was the average number of copies per run?		
d. How much money did we save this week?		

EXPLORATION 3.21   **Making Sense of Little Numbers That Add Up**

1. Let's say someone has a leaky faucet. How much does that drip add up to?

    a. First, discuss and note any assumptions that you will make in doing the problem.
    b. Determine your answer.

2. Environmental scientists need a lot of information to determine how pollution is affecting the environment. For example, trees give off oxygen, and the amount of oxygen given off is proportional to the total area of all the leaves on a tree. Thus, they need to estimate the number of leaves on a tree and the area of an average leaf. Let us tackle a simpler problem here: How many blades of grass do you think there are in an average lawn? Write down your guess.

    a. Let's make the question more focused now. Your class will obtain a piece of sod 1 foot by 1 foot. How many blades of grass do you think there are in this sample? Write your guess and any reasoning that you did.
    b. Share your guess with other members of your group or class. If you wish to change your guess now, do so and briefly explain what ideas caused you to change.
    c. How might you determine the number of blades of grass without counting each one? Write down your ideas and justification.
    d. Share your plans with other members of your group or class. At this point, each group will design and carry out its plan. Describe and justify your plan; then carry it out. Show your work.
    e. Each group will present its plan and "answer" to the whole class.
    f. If all answers were close, fine. If not, which answer do you feel is closest to the actual amount? Why?

Number Theory

Many people do not realize that mathematical concepts and ideas often have visual representations. For example, in Chapter 3 you investigated different visual representations of multiplication. In this chapter's explorations, you will be able to visualize some of the concepts that we will study in the text. These explorations can be done with elementary school children, yet they offer interesting challenges to the adult student too.

In each of the following explorations, you will not only lay the foundation for learning some important number theory ideas but will also have the opportunity to develop more mathematical tools—making and testing predictions. Students often tell me that when they think of mathematics, they think of numbers and computations. However, making and testing predictions has been crucial not only to the development of mathematics but also to the development of civilization. Young children are constantly making and testing predictions; I have cleaned up many messes my children made as a result of erroneous predictions! People make and test predictions regularly in their personal and professional lives. Yet somehow, this important human activity seems to be absent from much of school mathematics.

SECTION 4.1

EXPLORING DIVISIBILITY AND RELATED CONCEPTS

As you discovered in Chapter 3, being able to decompose and recompose a number in different ways is crucial to understanding computation algorithms for the four operations and to doing mental arithmetic and estimation confidently. Depending on the situation, we might decompose 48 as $40 + 8$, as $50 - 2$, as $8 \cdot 6$, or as $2 \cdot 2 \cdot 2 \cdot 2 \cdot 3$. In the following exploration, as you play the Taxman game more times, you will discover that one kind of decomposition of a number will enable you to increase your score, and you will see new relationships among numbers.

EXPLORATION 4.1   **Taxman**

I don't know who invented this game, but I have seen variations in many different places. The version I present to you is one in which a team of two (or more) players competes against the "taxman."

Materials

■ Array of consecutive numbers, such as 1 through 20 (see the game sheets on pages 107 and 109).

Rules

1. The team crosses out a number.

2. The taxman crosses out all the proper factors of that number.

3. The team crosses out another number. A number can be crossed out only if at least one of its proper factors has not yet been crossed out. For example, you can't cross out 14 if 1, 2, and 7 have all been crossed out already. Play continues until none of the remaining numbers has a proper factor that has not been crossed out.

4. At the end of the game, the taxman gets all remaining numbers!

5. To determine your score, add up all the numbers your team crossed out.

1	2	3	4	5	6	7	8	9	10
11	12	13	14	15	16	17	18	19	20

1.   Take out the Taxman Game Sheet on page 107. Play several games, using the numbers 1 through 20.

   a. In the table provided on the game sheet, record your best game. Briefly justify the choice for each number.

   b. Describe and briefly justify any general strategies that you think will be useful when playing the game with bigger numbers.

   c. After listening to the class discussion, if your game was not the highest-scoring one, describe what you learned from the class discussion.

2.   Take out the Taxman Game Sheet on page 109. Play several games using the numbers 1 through 30.

   a. In the table provided on the game sheet, record your best game. Briefly justify your choice for each number.

   b. Describe and briefly justify any general strategies that you think will be useful when playing the game with bigger numbers.

   c. After listening to the class discussion, if your game was not the highest-scoring one, describe what you learned from the class discussion.

3.   The following activity is designed to help you develop communication skills and to see the extent to which you can zero in on the most fundamental relationships in the game. Write a set of directions for an imaginary friend Becky, who knows the rules for the game and knows basic mathematics vocabulary but has never played the game. That is, write a set of directions that are general enough so that she can use them no matter how many numbers she is using.

4.   If you were going to play a game from 1 to 100, describe your first three turns and explain your choices.

Taxman Game Sheet for EXPLORATION 4.1, Step 1

1	2	3	4	5	6	7	8	9	10	
11	12	13	14	15	16	17	18	19	20	Score ____

1	2	3	4	5	6	7	8	9	10	
11	12	13	14	15	16	17	18	19	20	Score ____

1	2	3	4	5	6	7	8	9	10	
11	12	13	14	15	16	17	18	19	20	Score ____

BEST GAME

Number	Justification

Taxman Game Sheet for EXPLORATION 4.1, **Step 2**

1	2	3	4	5	6	7	8	9	10

11	12	13	14	15	16	17	18	19	20

21	22	23	24	25	26	27	28	29	30

Score ____

1	2	3	4	5	6	7	8	9	10

11	12	13	14	15	16	17	18	19	20

21	22	23	24	25	26	27	28	29	30

Score ____

1	2	3	4	5	6	7	8	9	10

11	12	13	14	15	16	17	18	19	20

21	22	23	24	25	26	27	28	29	30

Score ____

BEST GAME

Number	Justification

SECTION   **4.2**   **EXPLORING PRIME AND COMPOSITE NUMBERS**

A full appreciation of the following explorations requires you to apply ideas of divisibility and enables you to appreciate the relevance of the concept of prime and composite numbers.

EXPLORATION 4.2   **Factors**

An exploration many elementary teachers use to help students develop number sense is to challenge them to make as many different rectangles as they can using a certain number of unit squares. For example, how many different rectangles can be made with six squares?

A lively discussion emerges, and one common debate question is whether there are four different rectangles or two different rectangles? What do you think?

Your answer depends on your definition of *different*. If we define *different* in terms of how the rectangle looks on the page, then we have four different rectangles. If we define *different* as having a different shape, then we have two different rectangles. This second definition is connected to the idea of congruence, which we will study later. In this case, we will use the second definition of *different*, not because it is "right" but because it connects to some concepts that we will soon develop.

We can make two different rectangles with six unit squares. If we represent them by their length and width, we have a 6 × 1 rectangle and a 3 × 2 rectangle. Using mathematical terminology, we can also say that 6 has these four factors: 1, 2, 3, 6.

When elementary school children investigate all the different rectangles they can make for each number, the exploration not only reinforces their multiplication facts but also addresses other important areas: problem-solving, communication, reasoning, and making connections. Because you are able to think at a more abstract level, we will modify the instructions given to the children.

1. Take out the Factors Table on page 113. In the table, list the factors of each of the first 25 natural numbers.
2. Describe any observations, hypotheses, and questions you have as a result of filling out and looking at the table.
3. One of the themes of this book is the power of different representations.
   a. Take out the Number of Factors Table on page 115. Use the data in your Factors Table from Step 1 to complete the table.
   b. Note any observations, hypotheses, and questions that you have as a result of filling out and looking at this table.
   c. In one sense, the numbers in each column of the table constitute a "family." Can you give a name to any of the families? If so, write down this name (it does not have to be a "mathematical" name) and describe the characteristics that are common to all members of the family.
   d. Suppose we were to continue to look at the factors of numbers beyond 25. Predict (and briefly describe your reasoning) the next number that will have 2 factors, the next number that will have 3 factors, and the next number that will have 4 factors.

e. Now extend the table until you have the next number that has 2 factors, 3 factors, and 4 factors. Describe any insights or discoveries from the extension.

f. Can you predict the first number that will have 7 factors? What will it look like? Justify your reasoning.

g. There are other possible families of numbers, based on the number of factors. For example, if we combine the 3-factor families, the 5-factor families, and the 7-factor families into a large family called "odd number of factors," how might you describe the characteristics that are common to all members of *this* family?

Factors Table for   EXPLORATION 4.2,   Step 1

Number	Factors							
1	1							
2	1	2						
3	1	3						
4								
5								
6								
7								
8								
9								
10								
11								
12								
13								
14								
15								
16								
17								
18								
19								
20								
21								
22								
23								
24								
25								

Number of Factors Table for EXPLORATION 4.2, **Step 3**

1 factor	2 factors	3 factors	4 factors	5 factors	6 factors	7 factors	8 factors
1	2	4					
	3						
	5						

EXPLORATION 4.3   **Finding All Factors of a Number**

In this exploration, we will be working with patterns (recognizing, describing, extending), making predictions and hypotheses, and using several related problem-solving tools, including being systematic and making tables.

1. Determine all the factors of 36. Show your work. Describe how confident you are that you have found *all* the factors and why you are very confident, somewhat confident, or not very confident.
2. Select some other numbers and find all of their factors. If you feel that your method is efficient, move on. If you don't feel that your method is very efficient, take some time to stop and think about how else you might determine all the factors of a number.
3. Describe your present method for finding all the factors of a number, as though you were talking to a classmate who missed this exploration.
4. Describe any other observations or hypotheses you have made up to this point.
5. Now compare your methods with those of your partner(s). If you like a method that you heard, summarize this new method in your own words.
6. How could you use the factors of 36 that you found in Step 1 to find all the factors of 72 without starting all over?
7. The number 200 has 12 factors: 1, 2, 4, 5, 8, 10, 20, 25, 40, 50, 100, and 200. This number 200 can be rewritten as $200 = 2^3 \cdot 5^2$. This representation is an example of a *prime factorization*, which is discussed in Section 4.2 of the text. Can you see any way to determine all 12 factors of 200 just from seeing this prime factorization?

SECTION **4.3** **EXPLORING GREATEST COMMON FACTOR AND LEAST COMMON MULTIPLE**

The following explorations, like the previous explorations in this chapter and like many other explorations in this volume, actually involve many concepts. If you did Exploration 4.1 (Taxman), you found that you dealt with divisibility in terms of finding factors of a number, and you found that your strategy for prime numbers was different from your strategy for composite numbers. So too with star patterns. In fact, you will encounter many number theory concepts in these explorations, and, as with the previous explorations in this chapter, your ability to decompose numbers and then analyze those decompositions and see relationships will enable you to unlock the secrets of star patterns—that is, to understand why they work.

EXPLORATION 4.4   Star Patterns

Star patterns have fascinated both mathematicians and students for hundreds of years. They are pleasing to the eye, and they contain many surprises. This exploration extends the work in the exploration with circle clocks in Chapter 3, in which our focus was on multiplication. As you make different star patterns, you will find that exploring relationships among numbers will help you to predict the shape of the figure and the number of paths and orbits needed to make a complete star pattern.

PART 1: Making sense of star patterns

Star patterns are made by beginning at some point on the circle and connecting points at regular intervals. For example, the star on the left below starts at point 0 and jumps to point 2, then to point 4, then to point 6, and then back to 0. The star on the right also begins at point 0, but it jumps three numbers at a time, and so it jumps from 0 to 3 to 6, then (3 points after 6) to 1, 4, and 7, then to 2 and 5, and finally back to 0. Depending on the number of points on the circle and the number of jumps between points, a variety of patterns are created!

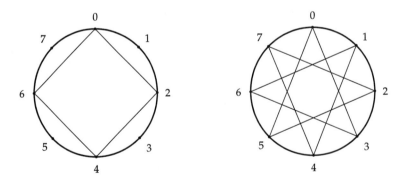

1. Take a minute to note what you see and think from these first two star patterns. Note your observations, questions, and hypotheses about these two demonstrations. Then share your thoughts with your partner(s). Add or modify any thoughts after the discussion.

2. *Patterns in the Stars*   The primary focus of this exploration is to examine the relationship between the number of dots we divide the circle into and the way we connect the dots. However, before we do that, we will focus on two other aspects of star patterns that reflect important mathematical ideas.

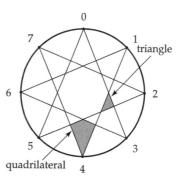

Let us stop to examine what we see in the stars. For example, look again at the eight-pointed star at the right. This figure has eight outer vertices, eight outer figures that are quadrilaterals, and eight smaller triangles.

a. What kind of quadrilateral is made? What kind of triangle is made? If you don't know the names, describe the characteristics of the quadrilateral and the triangle.

b. What else do you see?

3. *Developing a Notation*   You have already seen that much of the mathematical notation that we use was invented quite recently. You have also seen that these notations are not facts but, rather, choices that people made because they thought that the new notation would make it easier to communicate mathematically. Because we will be exploring star patterns, it would be nice to have a simpler way to refer to different star patterns than, for example, "Divide a circle into 8 arcs and then connect every fifth dot." Write down your ideas for ways of representing different star patterns that would be shorter than the present verbal descriptions.

4. *Initial Hypothesis*   Look again at the star patterns on p. 118. We will refer to the star pattern on the left as star (8, 2) and to the pattern on the right as star (8, 3). This common notation for describing star patterns uses the ordered-pair idea developed in Chapter 2: The first number tells us how many equal arcs the circle has been divided into, and the second number tells us the rule for connecting dots. However, other notations for naming star patterns are possible, such as S_2^8 and S_3^8; that is, this way of naming star patterns is a **convention.**

Now let us focus on the relationships between these two numbers—that is, the number of divisions of the circle and how the points are connected. Looking back to our first two star patterns, which we will now call star patterns (8, 2) and (8, 3), we found that star pattern (8, 2) is a square and star pattern (8, 3) is an eight-pointed star. The question that we will now explore is: Just from looking at the two numbers in any star pattern (x, n), can you predict what the star pattern will look like?

Brainstorm with your group how you might explore this question. What strategies seem more reasonable or potentially productive? On the next page are several circles that you can trace and then see what patterns emerge when you use different values for x and n. After brainstorming and preliminary exploration, answer the following questions.

a. Describe the strategies that you chose to employ and briefly explain your reasons for choosing those strategies.

b. Describe your observations, the questions that occurred to you, and any preliminary hypotheses you formed on the basis of your initial explorations. Explain why you think your hypotheses are true. Note that in mathematics, *true* means true 100 percent of the time. If a hypothesis is true 99 percent of the time, we say that it is false.

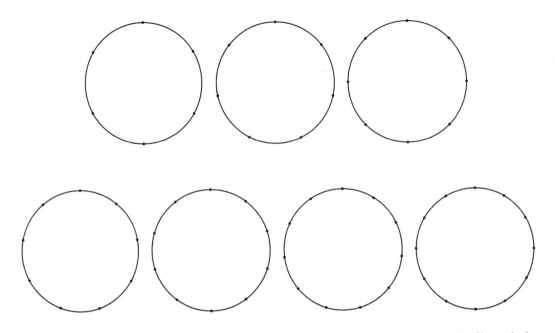

The following activities will help you to generate more data systematically to help you test and refine your hypothesis from Part 1.

PART 2: From observations to hypotheses to generalizations

Exploring One Star Family

In exploring a phenomenon, if often helps to be systematic. This is true not only in mathematics. For example, much medical research involves systematically exploring different combinations of drugs; archaeological work is very systematic; and much research in astronomy involves a systematic examination of data.

In this case, we will systematically examine all members of the (8, n) star family—that is, all seven star patterns that can be made by dividing a circle into eight arcs.

1. Take out the star families on page 123. Make the stars in the (8, n) star family and record your thinking and your work according to the following directions. In each case, first predict what you think the star will look like. Afterwards, if your prediction was correct, move on. If it was not, reflect on what happened, explain why your prediction was not accurate, or explain what you learned in drawing the star.

2. After making all of the patterns, do the following:

   a. Describe any observations that you made and try to explain those observations. For example, patterns (8, 3) and (8, 5) are the same. Can you explain why?

   b. State any hypotheses you have about star patterns in general. For example, can you make a prediction about what happens when the second number does or does not divide the first number evenly?

   c. Note any questions that you have.

Exploring Another Star Family

3. Make the stars in the (7, n) star family. As before, first predict next what you think you will find, next make the pattern, and then reflect on your prediction if it was not correct.

4.   After making all the patterns, do the following:
   a.   Describe any observations that you made and try to explain those observations.
   b.   State any hypotheses you have about star patterns in general.
   c.   Note any questions that you have.

Gathering More Data

In its most open-ended form, the task of this exploration is to note what you observe about star patterns in order to predict a star's appearance simply from knowing the number of points on the star and the number of jumps. That is, for star (x, n), if you are given the values of x and n, can you predict what it will look like?

5.   Because so many students are so new to this idea of looking for patterns and making and testing predictions, the directions here will not be quite so open-ended. Therefore, make other star families on your own — for example, circles with 6 dots, 9 dots, 10 dots, and so forth (Use the star families provided on pages 125 and 127). As you do so, look for patterns. For example, depending on the x and n, some stars will look like geometric figures (triangles, squares, and so forth), and some stars will look very similar to other stars.

6.   When making star patterns of the form $(8, n)$ and $(7, n)$, you found that some of the stars were equal; for example, star $(8, 3) =$ star $(8, 5)$. Can you jump from this observation to a more general hypothesis? For example, when will $(x, a) = (x, b)$?

7.   Some of you may have realized that there are other instances in which two star patterns will be equal. For example, draw the $(12, n)$ star family and the $(6, n)$ family if you have not yet done so. Is it possible that there are numbers a and b such that star $(12, a) =$ star $(6, b)$? Can you state your observations as one or more hypotheses?

8.   What other hypotheses have you come up with? State your hypotheses and explain why you think they are true.

9.   The following questions that are often raised by students deal with the ideas of limitations on x and n. Answer these questions.
   a.   Is it possible to have star (x, x)? If so, what would it look like? If not, why not?
   b.   Is it possible to have star $(x, 0)$? If so, what would it look like? If not, why not?
   c.   Is it possible to have star $(x, {}^-1)$? If so, what would it look like? If not, why not?

Star Families for EXPLORATION 4.4, **PART 2, Steps 1 and 3**

1. $(8, n)$ **star family**

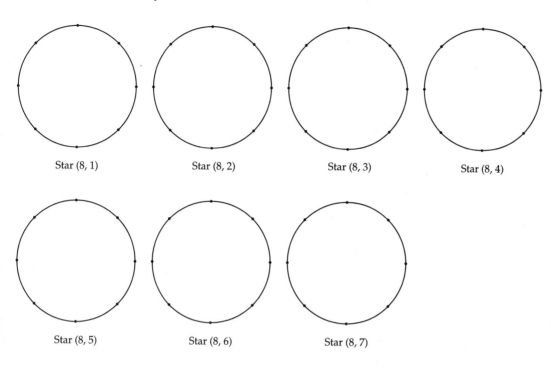

| Star (8, 1) | Star (8, 2) | Star (8, 3) | Star (8, 4) |

| Star (8, 5) | Star (8, 6) | Star (8, 7) |

3. $(7, n)$ **star family**

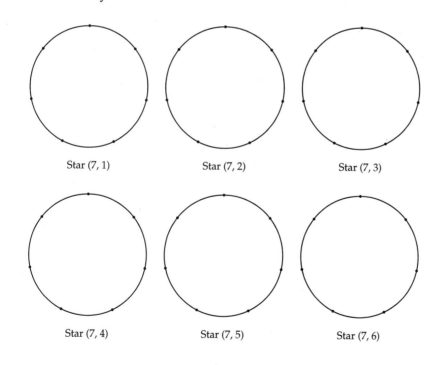

| Star (7, 1) | Star (7, 2) | Star (7, 3) |

| Star (7, 4) | Star (7, 5) | Star (7, 6) |

Star Families for   EXPLORATION 4.4,   **PART 2, Step 5**

(6, *n*) star family

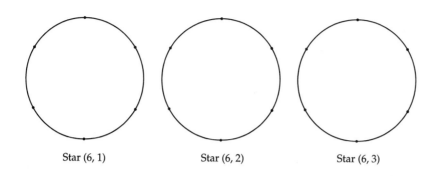

Star (6, 1)   Star (6, 2)   Star (6, 3)

(9, *n*) star family

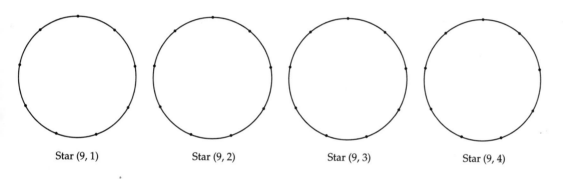

Star (9, 1)   Star (9, 2)   Star (9, 3)   Star (9, 4)

(10, *n*) star family

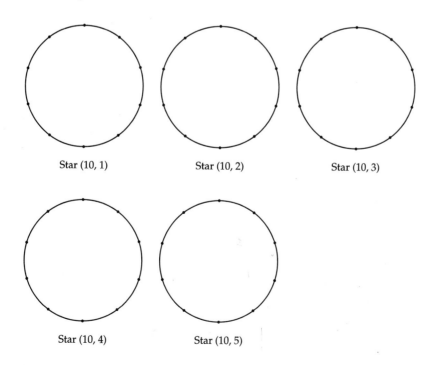

Star (10, 1)   Star (10, 2)   Star (10, 3)

Star (10, 4)   Star (10, 5)

Star Families for   EXPLORATION 4.4   PART 2, Step 5

(11, _n_) star family

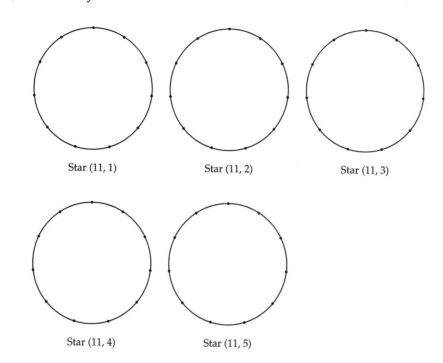

Star (11, 1)   Star (11, 2)   Star (11, 3)

Star (11, 4)   Star (11, 5)

(12, _n_) star family

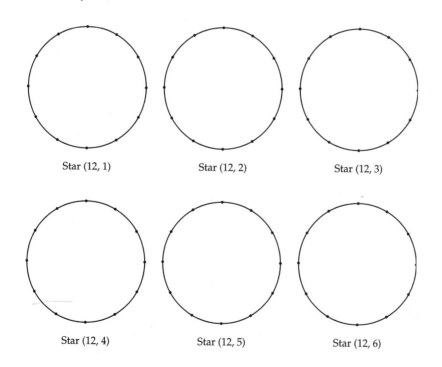

Star (12, 1)   Star (12, 2)   Star (12, 3)

Star (12, 4)   Star (12, 5)   Star (12, 6)

5

Extending the Number System

M any beginning elementary teachers think that they will never have to teach fractions, integers, or decimals unless they teach upper elementary school. This may have been true in the past, but it is not true anymore. Even in kindergarten, children encounter the idea of fractions in terms of dividing. For example, "We each get half." Second graders will often ask, "Why can't you take 8 from 2?" when a teacher introduces regrouping (for example, $52 - 38$). Long before formal work with decimal operations, children encounter decimals when they deal with money. Therefore, it is essential that all elementary teachers have an understanding of the relationship of these three systems (integers, fractions, and decimals) to whole numbers and to each other.

SECTION 5.1 EXPLORING INTEGERS

Although most students do not encounter integers until middle school, many elementary school children are aware of them, and quite a few will be able to work with them. Therefore, it is important that the elementary teacher be comfortable with integers from a conceptual point of view—that is, how they connect to positive whole numbers and how operations with integers are like and unlike operations with positive whole numbers. The following explorations are designed to help you better understand integer operations.

EXPLORATION 5.1   **Understanding Integer Addition**

When we examined addition with positive whole numbers, one model we used to represent this operation was the set (discrete) model. Let us use this model to develop rules for integer addition. We will use black dots to represent positive numbers and white dots to represent negative numbers.[1] Thus, the figure below is a representation of 3 + 4 in the context of addition as joining. That is, we have two sets, one containing 3 dots and the other containing 4 dots. When we combine the two sets, we have 7 dots; that is, 3 + 4 = 7.

1. Model the problem 5 + ⁻2. Describe and justify your process so that a reader who is not familiar with integer addition could understand both how you got the answer and why your process works.
2. Model the problem ⁻7 + 5. Describe and justify your process so that a reader who is not familiar with integer addition could understand both how you got the answer and why your process works.
3. Model the problem ⁻3 + ⁻6. Describe and justify your process so that a reader who is not familiar with integer addition could understand both how you got the answer and why your process works.
4. We have now examined all four cases that can occur when adding integers: both numbers positive, the larger magnitude positive, the larger magnitude negative, and both numbers negative. Based on your work above, can you describe a general rule for adding integers that will work for all cases?

---

[1] In many elementary classrooms, teachers will use black chips to represent positive numbers and red chips to represent negative numbers.

EXPLORATION 5.2   **Understanding Integer Subtraction**

As you discovered in Chapter 3, different contexts for subtraction lead to very different diagrams. We will examine integer subtraction first in the take-away context and then in the comparison context.

Subtraction as Take-Away

A representation of $5 - 2$ from the take-away perspective follows. That is, the figure at the left below shows 5 objects. The figure at the right shows us taking 2 away, and thus 3 objects remain; that is, $5 - 2 = 3$.

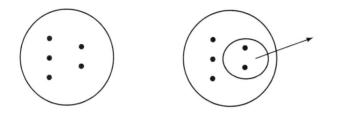

1. Model the problem $2 - 5$; that is, 2 take away 5.
2. Describe and justify your process so that a reader who is not familiar with integer subtraction could understand both how you got the answer and why your process works.
3. Many students struggle to find a way to model this problem. As with all mathematics, someone, somewhere, sometime was able to come up with a model that was useful. The key to the following model is to use the idea of *additive inverses:* For any number a, there is a number ^-a (the additive inverse) such that $a + {}^-a = 0$.
    The circle to the right represents the first step in the problem $2 - 5$. Do you see why? Work on this before reading on. . . .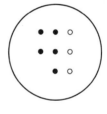
        The overall value of the objects in this circle is still $^+2$. (Recall the standard algorithm for subtracting whole numbers when regrouping is involved. We change the representation of the minuend—the top number—but we don't change its overall value.) Now we can apply our take-away model of subtraction. We can "take away" 5; that is, we take away 5 black circles. Thus, we have demonstrated with the chips that $2 - 5 = {}^-3$, because the value of what is left is $^-3$.
4. Model the problem $^-3 - 6$. Describe and justify your process so that a reader who is not familiar with integer subtraction could understand both how you got the answer and why your process works.
5. Model the problem $4 - {}^-5$. Describe and justify your process so that a reader who is not familiar with integer subtraction could understand both how you got the answer and why your process works.
6. Model the problem $^-2 - {}^-4$. Describe and justify your process so that a reader who is not familiar with integer subtraction could understand both how you got the answer and why your process works.

7. We have now examined all four cases that can occur when we are subtracting integers using the take-away model. On the basis of your work above, can you describe a general rule for subtracting integers that will work for subtraction as take-away?

Subtraction as Comparison

As you saw in Chapter 3, and as most first and second graders will readily tell you, subtraction as comparison is quite different from subtraction as take-away. The key to being able to model integer subtraction resulting from the subtraction-as-comparison context is to find a way to adapt our notion of subtraction as comparison so that it works with integers. For example, we could describe $6 - 2$ as comparison by saying that this means "How much bigger is 6 than 2?" See the figure at the right. We can readily see that 6 is 4 bigger than 2. However, this way of describing subtraction as comparison doesn't help us with problems like $^-3 - 6$, because we would ask, "How much bigger is $^-3$ than 6?"

8. Work with your partners to find a way to describe subtraction as comparison that will work for the cases we examined with take-away: $2 - 5$, $^-3 - 6$, $4 - ^-5$, and $^-2 - ^-4$.
   a. Model the problem $2 - 5$. Describe and justify your process so that a reader who is not familiar with integer subtraction could understand both how you got the answer in the context of subtraction as comparison and why your process works.
   b. Model the problem $^-3 - 6$. Describe and justify your process so that a reader who is not familiar with integer subtraction could understand both how you got the answer in the context of subtraction as comparison and why your process works.
   c. Model the problem $4 - ^-5$. Describe and justify your process so that a reader who is not familiar with integer subtraction could understand both how you got the answer in the context of subtraction as comparison and why your process works.
   d. Model the problem $^-2 - ^-4$. Describe and justify your process so that a reader who is not familiar with integer subtraction could understand both how you got the answer in the context of subtraction as comparison and why your process works.

9. We have now examined all four cases that can occur when we are subtracting integers using the comparison model. On the basis of your work above, can you describe a general rule for subtracting that will work for subtraction as comparison?

Looking Back

On the basis of your work above, can you describe and justify a general rule or set of rules for *any* subtraction problem?

EXPLORATION 5.3   **Understanding Integer Multiplication**

Below is a representation of 4 · 3 using chips. It uses the repeated addition context for multiplication. That is, we have four sets of 3, or three 4 times.

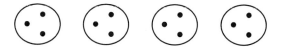

1.   What about 4 · ⁻3?

    a.   Model and solve this problem by using or adapting the idea of multiplication as repeated addition.

    b.   Describe and justify your process so that a reader who is not familiar with integer multiplication could understand both how you got the answer and why your process works.

2.   What about ⁻4 · 3?

    a.   Model and solve this problem by using or adapting the idea of multiplication as repeated addition.

    b.   Describe and justify your process so that a reader who is not familiar with integer multiplication could understand both how you got the answer and why your process works.

3.   Model the problem ⁻4 · ⁻3. Describe and justify your process so that a reader who is not familiar with integer multiplication could understand both how you got the answer and why your process works.

4.   We have now examined all four cases that can occur when we are multiplying integers. On the basis of your work above, can you describe a general rule for multiplying integers?

EXPLORATION 5.4   **Understanding Integer Division**

Repeated Subtraction and Partitioning Models

Consider the problem $8 \div 2$. We learned in Chapter 3 that we can represent this problem using the partitioning model of division, and we can represent this problem using the repeated subtraction model of division. For review, you may want to model the problem $8 \div 2$ with the partitioning model and with the repeated subtraction model.

1. **a.** Model the problem $^-8 \div 2$ with the partitioning model.
    b. Model the same problem with the repeated subtraction model.
    c. Describe and justify your process so that a reader who is not familiar with integer division could understand both how you got the answer and why your process works.

Missing Factor Model

When we get to other cases of integer division (for example, $8 \div ^-2$ and $^-8 \div ^-2$), these two models become more and more complex, and at some point, they pass into the realm of "more trouble than it's worth." Therefore, let us abandon the partitioning and repeated subtraction models for division and look to another model for division to justify the procedures for integer division, the missing factor model. When we use this model, we find the solution to $a \div b$ by asking, "What number times b is equal to a?"

2. Use the missing factor model to determine the answers to the following three problems. As before, describe and justify your process so that a reader who is not familiar with integer division could understand both how you got the answer and why your process works.

$$^-8 \div 2$$
$$12 \div ^-3$$
$$^-20 \div ^-4$$

3. We have now examined all four cases that can occur when dividing integers. On the basis of your work above, can you describe a general rule for dividing integers?

SECTION ◆ **5.2** ◆ **EXPLORING FRACTIONS AND RATIONAL NUMBERS**

Researchers have told us for years that students' understanding of rational numbers is poor. This is unfortunate because the concept of rational numbers is one of the "big ideas" in elementary school mathematics. The fraction explorations here have been designed to have you work with the fundamental concepts related to fractions. Just as we discussed different decompositions of whole numbers in Chapters 3 and 4, a key to understanding fractions is to decompose them. That is, the notion of parts and wholes connects to our compositions and decompositions with whole numbers. We shall investigate various decompositions, which in turn will deepen your understanding of the different fraction contexts and the relationship between the numerator and the denominator.

As you found with whole numbers, one can be able to compute but still not understand the operations and procedures. As you develop your understanding of the many ways in which fractions can be interpreted and how these interpretations connect to the meanings of addition, subtraction, multiplication, and division, you will be able to work with fractions confidently and powerfully—that is, you will be able to solve multistep and nonroutine problems.

EXPLORATION 5.5   **Making Manipulatives**

Using manipulatives in elementary school mathematics was one of the bandwagons of the 1980s. In fact, "hands-on" became one of those infamous buzzwords. However, hands-on alone is not enough. One modification of this phrase is "hands-on and minds-on." In other words, although manipulatives can help to ground one's understanding of new ideas, what the hands do needs to be connected to important ideas in the mathematical concept.

One of the reasons for the limited success of much work with manipulatives on fractions is that even when students have worked with manipulatives, they have often worked with premade manipulatives. Although there is a place for premade manipulatives in the curriculum, many mathematics educators believe that at some point early in their development, students will benefit tremendously by making their own manipulatives. Such an activity not only allows the students to grapple with concepts at a deeper level but also encourages more creativity on their part. The following exploration has been one of my students' favorites.

Making the Manipulatives

Some of you will make fraction manipulatives from circles and some from squares. Regardless of which manipulatives you make, read the questions in Steps 1–4 below and keep them in mind as you make your manipulatives.

Circles: Using construction paper that has been cut into circles of the same size, make a set of manipulatives. The only restriction is that you cannot use a protractor. For example, you cannot make 1/4 circles by measuring 90-degree angles. In addition to physical tools, you will also need to use a combination of problem-solving strategies, including reasoning and guess–check–revise.

Squares: Using construction paper that has been cut into squares of the same size, make a set of manipulatives. The only restriction is that you cannot use a ruler to measure the divisions, although you can use a ruler to draw straight lines. For example, if your square is 6 inches on a side, you cannot use the ruler to make a mark every 2 inches. In addition to physical tools, you will also need to use a combination of problem-solving strategies, including reasoning and guess–test–revise.

After you have made your manipulatives, respond to the following questions.

1.   Describe at least one strategy that you used when making the manipulatives—that is, a strategy that you used for more than just one or two denominators.

2.   **a.**   Describe how you made thirds as though you were talking to someone who missed class. Your description needs to have enough specificity so that the person could repeat what you did and see why you did it that way.

   b.   Describe how you made fifths as though you were talking to someone who missed class.

3.   Describe at least two learnings that resulted from making your set of manipulatives. These descriptions may be framed as statements or hypotheses.

4.   Describe any questions that you have at this point, either questions about how to make a particular fraction (such as ninths) or other questions that arose when you were making your sets.

5.   Imagine explaining fractions to an Alphabitian.

   a.   How would you define the term *fraction* in your own words?

   b.   How would you define the term *numerator* in your own words?

   c.   How would you define the term *denominator* in your own words?

6. Alicia said that making 1/5 was hard. Brandon said that it was easy because 1/5 is "halfway between 1/4 and 1/6." Jamie said that Brandon's method can be used for lots of cases; for example, 1/7 will be halfway between 1/6 and 1/8, and 1/6 will be halfway between 1/4 and 1/8.

    a. What is your initial reaction to Brandon's statement? Do you feel that 1/5 is halfway between 1/4 and 1/6 or not? Why?

    b. Discuss this question in your group. If you changed your mind, explain what changed your mind and justify your present position. If you didn't change your mind, but you feel you can better justify your response, write your revised justification below.

7. Leah called some fractions "prime fractions." What do you think she meant?

EXPLORATION 5.6 **Ordering Fractions**[2]

With the widespread use of calculators and the increasing use of the metric system, the use of fractions in computation in everyday life has diminished considerably. However, understanding fraction relationships is still important because of the connection between fractions and ratios and the connection between fractions and decimals. Understanding these fraction relationships is an important underpinning of what we call *proportional reasoning*.

In this exploration, you will be asked to use your knowledge of fractions and your reasoning to order fractions. We will ask you to do this *without finding the LCM or using a calculator to convert them to decimals* because the goal here is not simply to get the right answer but to apply basic fraction ideas. Similarly, although you are encouraged to use the manipulatives you made in Exploration 5.5 to help you answer the questions, the justification needs to go beyond drawing a picture of the two fractions and saying that one looks bigger. For example, 5/9 is greater than 4/9 because both fractions have the same denominator, and hence each of the pieces is the same size. Therefore, having 5 such pieces is more than having 4 such pieces. As you grapple with determining which fraction in each pair is greater, see what connections you can make to your understanding of the terms *fraction, numerator,* and *denominator*.

Take out the tables on pages 139–140, for use in Steps 1 and 4 below.

1. Use the table on page 139. Predict the relative values of the fractions by inserting the symbol $<$, $=$, or $>$ into the space between the fractions, using fraction ideas without finding the LCM or converting to decimals. Briefly justify your choice.
2. Compare your responses with your partner(s). In some cases you may have the same answer but different justifications. Listen to one another; remember that just because you have the right answer does not necessarily mean your reasoning is valid. By the same token, a wrong answer often contains some valid reasoning. On some questions, you may disagree. Listen to one another; see whether you can find the misconception(s) in someone's reasoning. After the discussion, take some time to write what you learned from the discussion.
3. With your partner(s), develop a set of general rules that could be used by another person. There are many possible rules and ways to state them. For example, if two fractions have the same denominator, the fraction with the larger numerator is greater. Justify at least two of your rules; that is, explain *why* they work.
4. Exchange rules with another group. Then use *the other group's rules* to order the fractions in the table on page 140.
5. Critique the other group's rules with respect first to validity and then to clarity. That is, if you feel that any of the rules are invalid, explain why. If you feel that any rules are not clear, circle the words or phrases that are ambiguous or unclear and explain why you find them so.
6. On the basis of the critique from the other group, make any necessary changes in your rules.
7. Order each set of fractions. Explain your thinking process.

a. $\dfrac{31}{80}$ $\dfrac{13}{17}$ $\dfrac{2}{3}$ b. $\dfrac{1}{3}$ $\dfrac{4}{7}$ $\dfrac{2}{5}$ $\dfrac{7}{8}$ $\dfrac{5}{16}$

c. $\dfrac{3}{10}$ $\dfrac{2}{3}$ $\dfrac{7}{12}$ $\dfrac{4}{5}$ $\dfrac{3}{7}$ d. $\dfrac{1}{8}$ $\dfrac{2}{5}$ $\dfrac{5}{8}$ $\dfrac{5}{6}$ $\dfrac{3}{49}$ $\dfrac{3}{56}$

[2] This exploration has been adapted from one developed by the Summermath for Teachers program at Mt. Holyoke College and the Mathematics for Tomorrow project at Education Development Center.

Table for EXPLORATION 5.6, Step 1

	Fraction	>, =, or <	Fraction	Justification
a.	$\dfrac{3}{5}$		$\dfrac{3}{8}$	
b.	$\dfrac{5}{6}$		$\dfrac{7}{8}$	
c.	$\dfrac{3}{5}$		$\dfrac{5}{12}$	
d.	$\dfrac{1}{2}$		$\dfrac{17}{31}$	
e.	$\dfrac{3}{8}$		$\dfrac{2}{9}$	
f.	$\dfrac{2}{7}$		$\dfrac{3}{8}$	
g.	$\dfrac{1}{4}$		$\dfrac{2}{9}$	
h.	$\dfrac{9}{11}$		$\dfrac{7}{9}$	
i.	$\dfrac{3}{8}$		$\dfrac{4}{10}$	
j.	$\dfrac{3}{10}$		$\dfrac{9}{23}$	

Table for EXPLORATION 5.6, Step 4

	Fraction	>, =, or <	Fraction	Justification
a.	$\dfrac{3}{4}$		$\dfrac{7}{12}$	
b.	$\dfrac{5}{8}$		$\dfrac{10}{13}$	
c.	$\dfrac{5}{12}$		$\dfrac{7}{13}$	
d.	$\dfrac{7}{10}$		$\dfrac{14}{19}$	
e.	$\dfrac{2}{7}$		$\dfrac{1}{3}$	
f.	$\dfrac{3}{8}$		$\dfrac{4}{7}$	

EXPLORATION 5.7   **Partitioning Wholes**

Let us briefly discuss the idea of multiple representations, which has become a cornerstone teaching principle for most mathematics educators. This idea is derived from the work of Jerome Bruner and Zoltan Dienes, and is discussed in Standard 10 of the NCTM Standards. As we discovered in Chapter 3, each of the four operations can be represented in different ways. With respect to fractions, it is crucial that you work with different representations. First, many mathematics educators have emphasized that students will construct much more powerful understandings of a concept if they explore that concept in a variety of physical contexts. Second, one representation is often more appropriate than others for a given problem situation.

There are many different ways in which we can represent fractions; in this text, we will focus on area models, discrete models, and number line models. Physical representations include circles, rectangles, pattern blocks, and geoboards for the area model; Cuisenaire rods and fraction bars for the number line model; and small circles for the discrete model. After a concept is introduced with one manipulative, many researchers recommend presenting the learner with a different manipulative; this will cause the learner to rethink the concept because most physical representations have different perceptual features that highlight some aspects of the concept(s) and do not address other aspects.[3]

FRACTION MODELS

Area Models		Discrete Models	Number Line Models
circles			
squares and rectangles		small circles OOOOO	Cuisenaire rods
pattern blocks			number line
geoboards			

The three parts of this exploration will ask you to work with these three different fraction models. The pages of the exploration are set up so that you can show all of your work in the space provided.

---

[3] Douglas T. Owens, ed., *Middle Grades Mathematics* (Reston, VA: National Council of Teachers of Mathematics, 1993), p. 122.

PART 1: Understanding area models

In each of the steps below, you will be asked to do the problems yourself and then to discuss your solution(s) with your partner(s). If you decide that some of your solutions were not correct, do not erase them. Describe your mistake, redo the problem, and justify your new solution.

1.   Suppose each of the rectangles below has the value of 1. Divide each rectangle into the number of equal pieces called for. Justify your solutions.

    a. 2 equal pieces

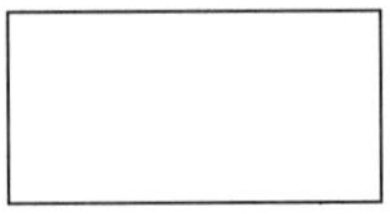

    2 equal pieces

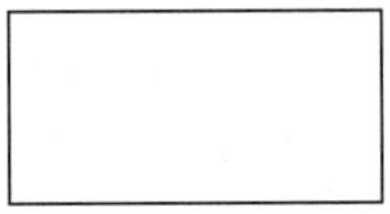

    b. 3 equal pieces

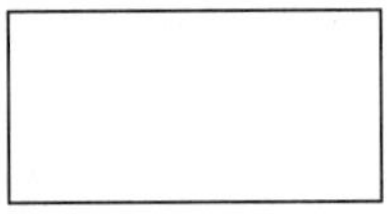

    3 equal pieces

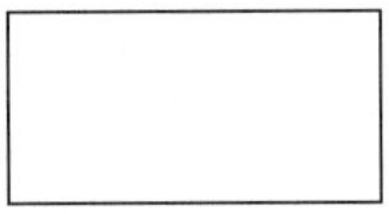

    c. 8 equal pieces

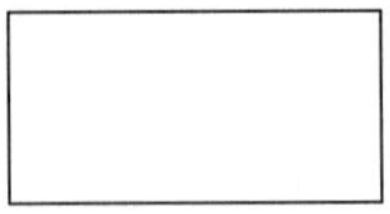

    8 equal pieces

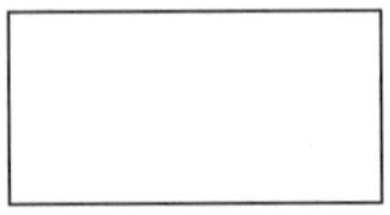

2.   Suppose each of the triangles below has the value of 1. Divide each triangle into the number of equal pieces called for. Justify your solutions.

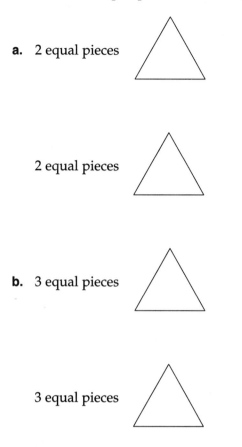

a. 2 equal pieces

2 equal pieces

b. 3 equal pieces

3 equal pieces

3.   Draw a diagram that has the specified value. Again, a primary purpose of these questions is to focus on the meaning of the numerator and denominator and on their relationship. Justify your solutions.

a. If the rectangle has a value of $\frac{1}{3}$, show 1.

b. If the rectangle has a value of $\frac{3}{4}$, show 1.

c. If the rectangle has a value of $1\frac{1}{3}$, show 1.

4. The following questions are related to pattern blocks, a manipulative commonly found in elementary school classrooms. Justify your solutions.

   a. If the hexagon has a value of $\frac{2}{3}$, show 1.

   b. If the trapezoid has a value of $\frac{3}{4}$, show 1.

   c. If the two hexagons together have a value of $\frac{4}{3}$, show 1.

   d. If the two hexagons together have a value of 1, show $\frac{1}{8}$.

   e. If the trapezoid has a value of $\frac{3}{4}$, what is the value of ▱?

5. **a.** If the whole region has a value of 1, what is the approximate value of the shaded region? Explain your reasoning.

b. Given that the large square has a value of 1, determine the approximate value of the small square. Explain your reasoning.

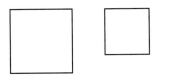

6. If each of the given hexagons below has a value of 1, does each individual piece in each of the hexagons represent $\frac{1}{6}$? Why or why not?

a.

b.

c.

7. Is the given figure a valid representation of $\frac{4}{3}$? Why or why not?

a.

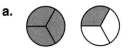

b.

PART 2: Understanding discrete models

1. Draw a diagram that has the specified value. Justify your solutions.

   a. If the given diagram has a value of $\frac{2}{3}$, show 1.

   • • •
   • • •

   b. If the given diagram has a value of 1, show $\frac{3}{4}$.

   c. If the given diagram has a value of $2\frac{2}{3}$, show 1.

   • • • •
   • • • •
   • • • •

   d. If the given diagram has a value of $\frac{1}{3}$, show $\frac{1}{2}$.

   • •
   • •

2. What fraction of the circles in the diagram below are black? Your answer needs to have a denominator other than 12. Justify your answer.

   ● ● ● ●
   ● ● ○ ○
   ● ● ○ ○

PART 3: Understanding number line models

1.   **a.**   On the number line below, mark the approximate location of 1. Explain your reasoning.

0         $\frac{1}{3}$

b.   On the number line below, mark the approximate location of 1. Explain your reasoning.

0                     $\frac{3}{4}$

c.   On the number line below, approximate the value of x. Explain your reasoning.

0   x 1

d.   On the number line below, approximate the value of x. Explain your reasoning.

0   x 6

Looking Back on Exploration 5.7

1. Which models were easier for you to work with? Why?
2. Which models were harder for you to work with? Why?
3. Describe your most important learnings.
4. Describe any problems that still puzzle you.

EXPLORATION 5.8   **Developing Fraction Sense**

After each part, compare your responses and your justification with those of your partner(s). As before, if you agree, listen to one another's justifications. If you disagree, discuss and debate until you reach agreement.

PART 1: Naming fractions[4]

For each step below, (a) give your answer, (b) explain your thinking (that is, how did you come up with your answer?), and (c) justify your response (that is, why do you believe it is correct?).

1. Name a fraction between 1/6 and 2/6.
2. Name a fraction between 9/10 and 1.
3. Name a fraction that is very close to 1. Now name a fraction that is even closer to 1.
4. Name a fraction that is greater than 1/2 but closer to 1/2 than to 5/8.
5. Give a value of x that makes the following statement true: $0 < \dfrac{4}{x} < \dfrac{1}{10}$.
6. Name a fraction between 0 and 1/10 that does not have a numerator of 1.

PART 2: Fraction benchmarks[5]

1. **a.** Place each of the fractions below in one of three groups: closer to 0, closer to 1/2, or closer to 1.

$$\frac{3}{8} \qquad \frac{2}{7} \qquad \frac{1}{3} \qquad \frac{21}{50} \qquad \frac{4}{5} \qquad \frac{7}{11} \qquad \frac{31}{181}$$

   b. Briefly justify each choice.
   c. Compare your responses and your justification with those of your partner(s). As before, if you agree, listen to one another's justifications. If you disagree, discuss and debate until you reach agreement.

2. **a.** Describe a general rule for deciding whether a fraction is closer to 0 or to 1/2.
   b. Describe a general rule for deciding whether a fraction is closer to 1/2 or to 1.

---

[4] Adapted from Activity 16 in Barbara Reyes, et al., *Developing Number Sense in the Middle Grades, Addenda Series, Grades 5–8* (Reston, VA: NCTM, 1991).

[5] Adapted from Activity 18 in Reyes et al., *Developing Number Sense in the Middle Grades.*

SECTION ◆ **5.3** **EXPLORING OPERATIONS WITH FRACTIONS**

As you may have already known before this course, being able to perform fraction computations and being able to apply fraction knowledge in real-life problems are not the same thing. As in Chapter 3, understanding the whys behind the whats is crucial if you are going to be an effective teacher of children. If you do the explorations in this section conscientiously, you will come out with a much stronger "fraction sense" and will better understand why the various fraction algorithms actually work!

EXPLORATION 5.9 **Developing Operation Sense**

This exploration focuses on further refining your operation sense, which we first explored in Chapter 3. This involves developing the ability to compute mentally with fractions and developing operation sense with fractions—that is, understanding the relative effect of different operations with fractions. Thus, please do not answer any of the questions by doing pencil-and-paper or calculator computations.

PART 1: Where is the answer on the number line?[6]

1. The points on the number line below are not drawn to scale. Without doing any computation, determine the region in which each answer will lie. Explain your reasoning.

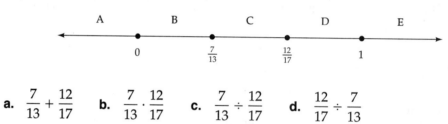

   a. $\dfrac{7}{13} + \dfrac{12}{17}$ **b.** $\dfrac{7}{13} \cdot \dfrac{12}{17}$ **c.** $\dfrac{7}{13} \div \dfrac{12}{17}$ **d.** $\dfrac{12}{17} \div \dfrac{7}{13}$

2. After discussing your answers and justifications with your partner(s), note important things that you learned from wrong answers, from different answers, or from correct answers. Try to connect these reflections to fraction ideas.

PART 2: Where do the digits go?

1. Using four different digits, make two proper fractions whose sum is as close as you can get to 1 but still less than 1.
2. Using four different digits, make two proper fractions whose difference is as great as possible but still a positive number.
3. Using four different digits, make two proper fractions whose product is as close as you can get to 1.
4. Using four different digits, make the smallest possible quotient.

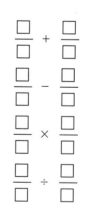

---

[6] This exploration has been adapted from one developed by the Summermath for Teachers program at Mt. Holyoke College and the Mathematics for Tomorrow project at Education Development Center.

EXPLORATION 5.10   **Wholes and Units: Not Always the Same**

Thinking of fractions only in terms of parts and wholes is simplistic and causes problems when we get beyond simple, routine problems. The following two problems will help you to grapple with some very important fraction ideas.

1. *Absent Students*[7]   Let's say you are teaching. It is winter, and it seems that a larger fraction of students than normal are sick. You are eating lunch in the teachers' lounge, and another teacher says, "In my class today, 2/5 of the girls were absent but only 1/5 of the boys were absent." What fraction of her class was absent today?

   a. Write down your initial thoughts about this problem.
   b. Meet with your partner(s) and discuss the problem. If you have the same answer, did you arrive at the answer in the same way? If you have different answers, listen to one another's reasoning until you can agree on one answer. Justify your answer.
   c. What is a realistic range of possible answers?
   d. What is the theoretical range of possible answers?

2. *How Much Is Her Share?*[8]   This problem is adapted from a real-life problem. Josephine is a graduate student at Urban State College. Because her financial resources are limited, she has moved into a house with four other people. The house is heated with electricity, and the electric bill comes every two months. Josephine moved in on February 1. When the bill for January–February comes, what fraction of the bill should each person pay?

   a. Work on this problem alone and show your work.
   b. What assumptions did you make in order to solve this problem?
   c. Meet with your partner(s) and discuss the problem. If you have the same answer, did you arrive at the answer in the same way? If you have different answers, listen to one another's reasoning until you can agree on one answer.

[7] This problem has been adapted from one developed by Deborah Schifter at Education Development Center.
[8] This problem has been adapted from a problem in *A Course Guide to Math 010L,* by Ron Narode, Deborah Schifter, and Jack Lochhead, 1985.

EXPLORATION 5.11   **Meanings of Operations with Fractions**[9]

The purpose of this exploration is to develop your operation sense by having you try to connect the problem with the operation.

PART 1: Matching problem situations with operations

For each of the seven problems below, do the following:

a.   Represent the problem with a diagram.
b.   Select the model that fits and briefly justify your choice:

+   combine, increase

−   take-away, comparison, missing addend

×   repeated addition, area, Cartesian product

÷   partitioning, repeated subtraction, missing factor

If none of the models fits, then explain how you knew which operation was appropriate.

c.   Write a number sentence that would answer the question, *but do not determine the answer.*

As an example, one solution for Problem 1 is modeled below.

1.   a.

b.   "At first, I did the problem by adding. That is, how many $\frac{3}{4}$ s can I add until I get to 12? But this didn't fit any of the models of the operations. Then I realized I could look at the diagram from "the other direction": How many $\frac{3}{4}$ s could I take away until I ran out of medicine? Then I realized this was repeated subtraction, and the appropriate number sentence popped into my head."

c. $12 \div \dfrac{3}{4}$

Problems

1.   A patient requires 3/4 of an ounce of medicine each day. If the bottle contains 12 ounces, how many days' supply does the patient have?

2.   The label on a bottle of juice says that 3/4 of the bottle consists of apple juice, 1/6 of the bottle consists of cherry juice, and the rest is water. What fraction of the bottle is juice?

3.   Freida had 12 inches of wire and cut pieces that were each 3/4 of an inch long. How many pieces does she have now?

4.   Jake had 12 cookies and ate 3/4 of them. How many cookies did he eat?

5.   Kareem had 12 gallons of ice cream in the freezer for his party. Last night Brad and Mary ate 3/4 of a gallon. How much ice cream is left?

6.   The Bassarear family is driving from home to a friend's house, and Emily and Josh are restless. They ask, "How far do we have to go?" Their father replies that

---

[9] This exploration is adapted from one developed by Ellen Davidson and Jim Hammerman at Education Development Center.

they have gone 12 miles and that they are 3/4 of the way there. What is the distance from home to the friend's house?

7.   Karla has 3/4 of an acre of land for her garden. She has divided this garden into 12 equal regions. What is the size of each region?

PART 2: Connecting whole-number and fraction contexts

1.   For each model that was not used in the seven problems in Part 1, write a story problem or explain why that model is not possible for fractions.

2.   Some of the models (such as the combining context for addition) apply to fractions in the same way that they apply to whole numbers. However, some of the models do not work in the same way. Determine which models do not work in the same way and explain why. For example, let's say you decided that the area model for multiplication of fractions works only if the fractions are proper fractions; explain why.

EXPLORATION 5.12   An Alternative Algorithm for Dividing Fractions[10]

The famous Indian mathematician Brahmagupta described an algorithm for dividing fractions that is actually easier to understand, in terms of *why* it works, than the traditional invert-and-multiply algorithm. The purpose of this exploration is to help you to understand this algorithm and why it works.

For each of the problems below, do the folowing:

a.   Represent the problem with a diagram and then solve the problem using the diagram.

b.   Describe what you did in order to arrive at your answer. It is important to consider carefully what you are actually doing, because careful thinking here increases the chances of the "breakthrough" in the next step.

c.   Try to connect what you did on paper to how the problem could be solved using only numbers. For example, in the first problem, the original number sentence is 4 ÷ 2/3, but regardless of the diagram you draw, you will divide each of your four units into thirds, and thus you are now solving 12/3 ÷ 2/3.

After completing the five problems, look for commonalities in all the problems that lead to a generalization (rule) that you could use in all the division problems. That is, you are trying to go from the concrete (pictorial) level to the abstract (numerical) level so that the manipulation of the numbers will make sense, just as you found for whole-number operations.

1.   Josie's Jammers have adopted a 4-mile stretch of highway to keep clean. Each afternoon they pick up trash. If they can clean 2/3 of a mile per day, how many days will it take them to clean the whole section?

2.   Chien has 2/3 of a gallon of gasoline, and each time he mows the lawn, he uses 1/6 of a gallon. How many times can he mow the lawn before buying more gasoline?

3.   Lyra is building dog houses. She has $7\frac{1}{2}$ pounds of nails. Each dog house requires $1\frac{1}{4}$ pounds of nails. How many dog houses can she make?

4.   Rita has $3\frac{3}{4}$ ounces of perfume and wants to sell the perfume in 3/8-ounce bottles. How many bottles of perfume can she make?

5.   Jonah has $3\frac{3}{5}$ pounds of dog food. Each day his dog eats 2/5 pounds. How many days' worth of dog food does he have?

---

[10] This exploration has benefited from the influence of Ellen Davidson and Jim Hammesman at Education Development Center.

EXPLORATION 5.13   **Remainders**

Recall the wire problem (Investigation 1.15). In many real-life situations, the remainder is as important as the quotient. The following exploration focuses on making sense of remainders with fractions.

Reconciling an Apparent Contradiction

1.   Marvin has 11 yards of cloth to make costumes for a play. Each costume requires $1\frac{1}{2}$ yards of material. How many costumes can he make?

    a.   Represent and solve the problem with a diagram.
    b.   Solve the problem using the division algorithm.

2.   Can you explain the fact that the diagram and the algorithm produce different answers?
    Note: Most students find this to be a rather challenging question. Discuss the question with your partner(s) before reading on.

Different Ways to Express Remainders

3.   Many students find going back and analyzing a simpler problem to be helpful. Consider the following problem: A doctor has 31 ounces of medicine, and each dose is 4 ounces. How many doses can the doctor make?

    a.   When doing the computation, we can represent the answer as $7\frac{3}{4}$ or as 7 R3.
    b.   What does the 3/4 mean?
    c.   What does the R3 mean?

4.   Now go back to Step 2. Write your present thinking about why we get two different answers.

5.   When we divide two fractions in which the quotient is not a whole number, sometimes we are interested in the fraction that the algorithm produces (for example, 1/3 in the Marvin problem). In other cases, we are interested in the fraction that the diagram produces (for example, 1/2 in the Marvin problem). Make up two story problems, one in which the fraction that the algorithm produces is relevant, and one in which the fraction that the diagram produces is relevant.

    Hint for Brahmagupta's algorithm: In order to answer each question, you need to subdivide the diagram further so that you can repeatedly subtract the second fraction (divisor).

SECTION **5.4**

EXPLORING BEYOND INTEGERS AND FRACTIONS: DECIMALS, EXPONENTS, AND REAL NUMBERS

There is much evidence that many students' understanding of decimals is mostly procedural. That is, they can compute with decimals, but their understanding of decimals is only weakly connected to their understanding of whole numbers and fractions and weakly connected to what the four fundamental operations mean. The following explorations will help you build stronger connections among many of the key concepts underlying operations with decimals.

EXPLORATION 5.14   Decimals and Base 10 Blocks

Connecting Decimals to Base 10 Blocks

For the following questions, you will need base 10 blocks.

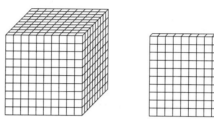

1. If the flat has a value of 1, represent 0.24 in two different ways. Justify your answers.
2. Represent 4.32 in two different ways. Specify your choice of unit for each representation.
3. Represent 0.0127 in two different ways. Justify your answers. Specify your choice of unit for each representation.
4. Demonstrate 0.463 + 0.507 with base 10 blocks. Justify each step.
5. Demonstrate 0.802 − 0.568 with base 10 blocks. Justify each step.
6. Show why $0.1 \times 0.1 = 0.01$ with base 10 blocks.
7. Demonstrate the equality of 0.1, 0.10, and 0.100 with base 10 blocks.

Greater Than or Less Than?

8. Insert < or > between each pair of numbers, and justify your choice:
    - **a.** 0.5   0.28
    - **b.** 8.3   8.14
    - **c.** 2.43   2.4168
    - **d.** 0.07   0.068
    - **e.** 6.74   6.74003

EXPLORATION 5.15   **Place Value: A Decimal Game**

Although most students can manipulate decimals (that is, add, subtract, multiply, and divide) with reasonable accuracy, there is much evidence that few students have a sophisticated number sense (such as being able to estimate) with respect to decimals. The NCTM recognizes that number sense is an essential tool for students to develop. The primary goal of each of the four games in Explorations 5.15 through 5.18 is for you to develop and refine your decimal number sense. It is important that you remember that the goal is not just to score the most points, but rather to monitor your own decimal sense constantly, asking questions such as:

What tools do I have to help me estimate decimal computations?

How is my ability to estimate related to my understanding of decimals?

The Game

This game requires you to pay attention to place value and to think about the relationship between the places. It can be played cooperatively or competitively.

Directions for playing the game

Step 1 Note the target number (for this game).

Step 2 Make a table like the one below. (*Note:* The places will vary depending on the places in the target number.)

Sum of the dice	10th	100th	1000th	Cumulative sum

Step 3 Roll two standard dice and write the sum of their values in one of the three place value columns. Exceptions: If the sum is 10, write a 0. If the sum is 11 or 12, write a 1.

Step 4 Record your cumulative sum in the last column.

Step 5 Repeat the last two steps until the game stops (explained below).

Step 6 The goal is to get as close as possible to the target number without going over the target number.

Step 7 The game stops when you go over the target number.

Here is an example of a record for one game with target number = 0.85:

Sum of the dice	10th	100th	Cumulative sum
6		6	0.06
8	8		0.86

1. Game 1: The target number is 8.00. The chart will have columns for ones, tenths, and hundredths.
2. Game 2: The target number is 78.42. The chart will have columns for tens, ones, tenths, and hundredths.
3. Make up your own target numbers and play more games.

Looking Back on Exploration 5.15

1. What if the new target number were 8.117? Describe and justify your overall "game plan."
2. Describe your most important learning from this game.

EXPLORATION 5.16   **The Right Bucket: A Decimal Game**

This game comes from *Ideas from the Arithmetic Teacher: Grades 6–8*,[11] one of many useful books published by the National Council of Teachers of Mathematics. It can be played alone or with a partner, competitively or cooperatively.

Materials

■ Game sheets on pages 161–163

Directions for playing the game

1. Select two decimals from the list given on the game sheet and cross them off the list.
2. Multiply the two decimals.
3. Determine your score by looking at the bucket chart below. For example, products between 10 and 100 earn a score of 2 points.
4 Select two more decimals, find their product, and determine your score. Continue until you have used all the decimals.
5. The goal is to score as many points as you can.

Bucket Chart

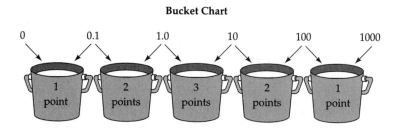

The learning value of this game comes from refining your ability to estimate. (You may want to reread the introduction to Exploration 5.15.)

1.   a. *First Game*   Use the game sheet on page 161. Select two decimals from the list provided, estimate their product, explain your reasoning, find the actual product, and determine your score. As you play, record your turns in the table provided on the game sheet. For example, the first turn in your game might look like this:

Turn	Decimals	Estimate	Reasoning	Actual product	Points
1	41.2 .083	4.12	0.083 is almost 1/10, and I knew 1/10 of 41.2 is 4.12. Thus, I felt that 0.083 of 41.2 would be well over 1.	3.4196	3

[11] George Immerzeel and Melvin Thomas eds., *Ideas from the Arithmetic Teacher: Grades 6–8* (Reston, VA: NCTM, 1982), p. 45.

   b. If you are playing competitively, record each person's scores for every turn and find each person's total score. The greatest total score wins. If you are playing cooperatively, record the team's ten scores and total score.

   c. What did you learn about estimating and mental math with decimals from this game?

2. **a.** *Second Game* Use the game sheet on page 162. Play the game over (a few times, if you wish), this time cooperatively. Your goal is to get the maximum number of total points.

   b. Is it possible to choose the pairs of decimals so that you can get 3 points each time? If you think so, explain why. If you think not, explain why not.

   c. Describe your maximum game in the table provided on the game sheet.

   d. Describe the different strategies that you used to determine which pairs of decimals to put together.

   e. What did you learn during the process of trying to find the maximum score?

3. **a.** *Third Game* Use the game sheet on page 163. Play another game with the new list of decimals. Record your game in the table provided.

   b. Describe any new strategies you devised from playing this game.

4. **a.** *Fourth Game* Make up a new set of decimals and determine the maximum score that can be obtained, using your decimals.

   b. Give your decimals to another team and have them determine the maximum score they can create.

The Right Bucket Game Sheet for   EXPLORATION 5.16,   Step 1

1.   a. *First Game*

| 3.6 | 0.03 | 13.1 | 29.6 | 11.9 | 0.7 | 33.7 | 0.04 | 21.9 | 0.125 |
| 10.1 | 0.42 | 0.07 | 2.9 | 0.29 | 19.5 | 5.52 | 23.1 | 9.6 | 1.8 |

Turn	Decimals	Estimate	Reasoning	Actual product	Points
1					
2					
3					
4					
5					
6					
7					
8					
9					
10					
				Total	

The Right Bucket Game Sheet for EXPLORATION 5.16, Step 2

2. a. *Second Game*

3.6	0.03	13.1	29.6	11.9	0.7	33.7	0.04	21.9	0.125
10.1	0.42	0.07	2.9	0.29	19.5	5.52	23.1	9.6	1.8

3.6	0.03	13.1	29.6	11.9	0.7	33.7	0.04	21.9	0.125
10.1	0.42	0.07	2.9	0.29	19.5	5.52	23.1	9.6	1.8

3.6	0.03	13.1	29.6	11.9	0.7	33.7	0.04	21.9	0.125
10.1	0.42	0.07	2.9	0.29	19.5	5.52	23.1	9.6	1.8

Turn	Decimals	Estimate	Reasoning	Actual product	Points
1					
2					
3					
4					
5					
6					
7					
8					
9					
10					
				Total	

The Right Bucket Game Sheet for EXPLORATION 5.16, **Step 3**

3. **a.** *Third Game*

| 0.023 | 0.9 | 5.74 | 0.245 | 7.3 | 1.2 | 23.8 | 16.5 |
| 0.068 | 8.7 | 0.12 | 1.8 | 0.78 | 42.7 | 0.2 | 3.6 |

Turn	Decimals	Estimate	Reasoning	Actual product	Points
1					
2					
3					
4					
5					
6					
7					
8					
9					
10					
				Total	

EXPLORATION 5.17   **Operations: A Decimal Game**

This game requires you to examine the effects of the four operations on decimal computation. As you move through the games, you should see improvement. At the end of each game, you will be asked to stop and reflect on your improvement.

Directions for playing the game

1. Make up four decimal numbers—for example, 3.6, 45.3, 1.23, 0.005.

2. From your list, select the three decimals that will make the answer to the given equation as large (or as small) as possible. Describe your strategy in words. For example, to make the equation

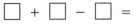

as large as possible using the decimals 3.6, 45.3, 1.23, and 0.005, you might try 45.3 + 3.6 − 0.005. Description of strategy: "I picked the two largest decimals to add and then picked the smallest decimal to subtract because I wanted to take away as little as possible."

3. If you are confident that you have the maximum number, go to Step 4. If not, try other combinations of numbers until you feel you have the combination that produces the largest number.

4. Make up four new decimal numbers. Give your strategy to another group. Have them use your strategy.
    Now check: Does this strategy produce the largest number?
    If yes, move on. If not, revise your strategy, and repeat Step 2.

1.   *First Game*   Make the answer to the equation below as large as possible.

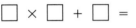

2.   *Second Game*   Make the answer to the equation below as *small* as possible.

3.   Would your strategy in the second game (Step 2) change if we added parentheses as shown below?

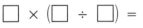

Describe your hypothesis before doing any computations with the calculator. Then do some computations and decide whether to keep your original strategy or to revise it.

4.   *Third Game*   This time you pick the operations, write the equation, and decide whether to make the answer to the equation as large or as small as possible.

Looking Back on Exploration 5.17

1.   Does your strategy depend on the numbers chosen? For example, for the sample exploration, we would pick the two largest numbers to put in the first two boxes and the smallest number for the third box, regardless of the numbers. Is there a simple, general strategy for every game?

EXPLORATION 5.18   **Target: A Decimal Game**

Materials

■ Game sheets on pages 167–168

Directions for playing the game

Note: These directions are written using an example that involves the operation of multiplication. The game can also by played using any of the other three operations: addition, subtraction, or division.

1. Select a number, an operation, a goal, and a winning zone; for example:

Starting number	145
Operation	multiplication
Goal	1
Winning zone	0.9 to 1.1 (or 0.99 to 1.01, or 1 to 1.1)

2. The first player selects a number to multiply the starting number by, trying to get the product of the two numbers in the winning zone (in this case, between 0.9 and 1.1).

3. If the product is not in the winning zone, then the product becomes the starting number for the second player.

4. Play continues in this manner until the product is within the winning zone.

1. Use the game sheets on pages 167–168. Play the game several times with a partner. Record your game in the table provided on the game sheet. For example, the first turn in the game described in the directions might look like this:

Turn	Computation	Product	Reasoning
1	145×0.006	0.87	I knew that 100×0.01 would be 1. Because 145 is bigger than 100, I picked a number smaller than 0.01.

2. Describe one mental math strategy that you learned during this game.

Target Game Sheet for   EXPLORATION 5.18

Starting number   _____

Operation   _____

Goal   _____

Winning zone   _____ to _____

Turn	Computation	Product	Reasoning
1			
2			
3			
4			
5			
6			
7			
8			

Target Game Sheet for EXPLORATION 5.18

Starting number   _____

Operation   _____

Goal   _____

Winning zone   _____ to _____

Turn	Computation	Product	Reasoning
1			
2			
3			
4			
5			
6			
7			
8			

Proportional Reasoning

The following explorations require you to grapple with ratios, rates, proportions, and percents, all of which fall under the "big idea" of proportional reasoning. It is ironic that although we probably find the need to use proportional reasoning in our everyday lives as much as we do any other kind of mathematical reasoning, a substantial body of research shows that few adults can competently and confidently navigate nonroutine problems involving proportional reasoning.

SECTION　6.1　**EXPLORING RATIO AND PROPORTION**

In this section, you will explore real-life applications of ratio and proportions, examine the connection between proportions and functions, and grapple with the meaning of ratios and proportions.

EXPLORATION 6.1   **Unit Pricing**

Unit pricing is now common. Underneath most items in grocery stores is the unit price of the item. For example, if a 24-ounce jar of pickles costs $2.79, the unit price is 11.6¢, which means that the pickles cost 11.6¢ per ounce, which is equivalent to $.116 per ounce.

If you don't see that 11.6¢ = $.116, STOP. Assess your understanding. Are you on solid ground, are you on shaky ground, or are you lost? If you are on shaky ground or lost, what can you do to connect to the material? What do you know that could help you understand? For instance, you might see that

$$10¢ = \$.10$$

So $$11¢ = \$.11$$

Thus $$11.6¢ = \$.116$$

Many states now require grocery stores to show the unit price below each item. One of the reasons for this law is that people tend to believe that larger items are proportionally cheaper. For example, they believe that a 64-ounce box of detergent will cost less than twice as much as a 32-ounce box. However, many companies use this belief to their advantage.

1.   Let's say you have Paul Bunyan pancake mix in two sizes: 30 ounces and 50 ounces. The smaller box costs $2.89, and the larger box costs $4.69.

   a.   Without using a calculator, determine which is the better buy. Record your process.

   b.   Use a calculator and record your process.

   c.   Jackie did the problem this way:

   $$\frac{30}{2.89} = 10.38 \qquad \frac{50}{4.69} = 10.66$$

   What labels belong with 10.38 and 10.66? In other words, what do those numbers mean?
   What do you think of Jackie's method? Justify your response.

   d.   Describe a situation in which you might buy the smaller item even if it cost more proportionally than the larger item.

EXPLORATION 6.2   **Buying Generic**

An issue related to unit pricing is generic products. In recent years, the sale of generic brands has increased tremendously. When I was growing up in the 1950s and 1960s, consumers could choose among the various name brands. Now, however, consumers can choose among the name brands or choose the generic alternative. Talk among your group. Are there products for which you are more likely to buy the generic brand? What about products for which you are not likely to buy the generic brand? What reasons do you give for choosing to buy the generic brand or choosing to buy the name brand?

1. One of the obvious reasons for buying the generic brand is that it is cheaper. In this exploration, we will examine how we can determine how much cheaper. Let's say you go to a store, and there you see a dispenser of Scotch brand tape that sells for $1.69 and a generic alternative that sells for $1.29. Both rolls contain the same amount of tape.

    a. How could we compare the two prices? Discuss this in your group and write down your thoughts.
    b. Describe and critique different ways in which we could compare the two prices.

2. Let's extend this question.

    a. Select a drugstore and collect and compare data on a name-brand pain reliever and a generic brand (or compare newspaper ads). Suppose you were writing an advertisement for the drugstore and you wanted to convince shoppers that they could save a lot of money by buying the generic pain reliever. Would your comparisons use addition or subtraction? (We will later call these comparisons *additive*.) Would they use multiplication, percents, or ratios? (We will later call these comparisons *multiplicative*.) Write the ad.
    b. Gather data on different sizes. Do you always get more for your money with the bigger size?
    c. Gather data from two different stores. How much cheaper is one store than the other?

3. Let's say an average family of four decided to buy generic whenever possible. Over the course of a year, how much money would they save?

EXPLORATION 6.3   **Proportional Reasoning and Functions**

In this exploration, we will explore a variety of functions that rely on proportional reasoning.

1.   Solve each of the problems below.

   a. It costs Rita 50¢ for the first 3 minutes of a long-distance call to her boyfriend and 20¢ for each additional minute. If Rita calls her boyfriend and talks for 12 minutes, how much does the call cost?

   b. At Sam's Submarine Sandwich Shop, the cost of your submarine sandwich is determined by the length of the sandwich. If Sam charges $2 per foot, how much will a 20-inch sandwich cost?

   c. The first-grade class is measuring the length of a dinosaur by using students' footsteps. The first dinosaur is 6 of Alisha's footsteps or 9 of Carlo's footsteps. If the second dinosaur is 8 of Alisha's footsteps, how many of Carlo's footsteps will it be?

   d. A farmer has determined that he has enough hay for 4 cows for 3 weeks. If the farmer suddenly obtains 2 more cows, how long can he expect the hay to last?

   e. Certain bacteria can double in number in 1 hour. If we start with 1 bacterium, how many bacteria will there be after 20 hours?

   f. Consider the set of quadrinumbers below. How many dots does the fifth quadrinumber contain?

2.   Discuss with your partner(s) solutions and strategies for each of the six problems in Step 1.

3.   Determine a general formula for each problem below. Then discuss solutions and strategies for each problem.

   a. If it costs 50 cents for the first 3 minutes and x cents for each additional minute, and a person talks for y minutes, what is the cost?

   b. If the sandwich costs x dollars per foot and it is y inches long, what is the cost?

   c. If the first dinosaur is 6 of Alisha's footsteps or 9 of Carlo's footsteps, and another dinosaur is x of Alisha's footsteps, how many of Carlo's footsteps will it be?

   d. If there is enough hay for 4 cows for 3 weeks, how long would the hay last if there were x cows?

   e. If bacteria double in number every hour, how many bacteria will there be after x hours? Start with 1 bacterium.

   f. How many dots are in the xth quadrinumber?

4.   For each problem, construct a graph that shows the relationship between the variables. Describe each graph as though you were talking to someone on the phone.

5.   Describe the similarities and differences you see among these six problems. For example, the graphs of some families are straight lines and the graphs of others are not.

EXPLORATION 6.4   **Safety First?**

Take a few minutes to examine the data below. Then respond to the questions that follow.

DEATHS FROM ACCIDENTS IN THE UNITED STATES

Year	Motor vehicles	Falls	Fires and burns	Drowning	Firearms	Drugs and medicine	Ingestion or inhalation of objects
1970	54,633	16,926	6718	6391	2406	2505	2753
1990	46,814	12,313	4175	3979	1416	4506	3303

1. What observations do you make from this set of data? If someone were to ask you what you see from these data, what would you say?
2. Describe at least two questions you might ask of the person who collected these data. One question needs to deal with the validity of the data.
3. The figures below show the same data, but in a slightly different form. How do you think the numbers in the second set of data were determined?

DEATH RATES FROM ACCIDENTS IN THE UNITED STATES
RATE PER 100,000 POPULATION

Year	Motor vehicles	Falls	Fires and burns	Drowning	Firearms	Drugs and medicine	Ingestion or inhalation of objects
1970	26.9	8.3	3.3	3.1	1.2	1.2	1.4
1990	18.8	5.0	1.7	1.6	0.6	1.8	1.3

4. Explain, as though you were talking to someone who doesn't understand, what the second set of numbers mean. For example, what does a motor vehicle death rate of 26.9 mean?
5. Let's say you were writing a newspaper article on changes in accidental deaths over this period and the editor said that because of space limitations, you could show either the raw numbers or the rates, but not both. Summarize the pros and cons of each representation.
6. Cecilia has a question. The number of deaths from ingestion or inhalation of objects rose from 2753 to 3303, and that is a 20% increase. How, then, can the rate have gone down?
7. Using only the data from the two tables, can you determine the U.S. population in 1970?
8. Select one column and predict whether you would expect the death rate to be higher or lower now than in 1990. Briefly explain your reasoning.
9. If you look at newspapers and magazines, you sometimes see raw data and sometimes see rates. For example, an article might discuss the number of births or the number of crimes, or it might discuss the birth rate or the crime rate. Explain, as though you were talking to someone who has not taken this course, the need for rates. What do rates mean? Why do we have them, and what purpose do they serve?

EXPLORATION 6.5   **Which Ramp Is Steeper?**

As we have discussed in the book, one of the uses of mathematics is to compare. In some cases, the comparison is easy—such as finding who is taller. In the following case, the comparison is not quite so simple. An exploration commonly done with elementary school students has them roll carts down ramps to examine the relationship between the steepness of the ramp and the distance the cart travels. In this exploration, we will focus on one aspect of this experiment—comparing the steepness of different ramps.

Suppose the students weren't told to standardize measurements, and each pair's ramp had different dimensions. For example, one pair had a ramp with a height of 14 cm, a base of 23 cm, and a ramp of 26.9 cm. Another pair had a ramp with a height of 12 cm, a base of 18 cm, and a ramp of 21.6 cm. One student hypothesized that the steeper the ramp, the further the marbles would go, and this prompted a desire to rank the slopes according to steepness. The students didn't want to build the ramps all over again, but each pair had the following measurements: length of base (*B*), height of base (*H*), and length of ramp (*R*).

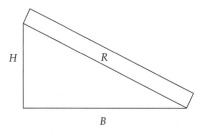

1. Develop a method for determining the relative steepness of the various ramps. *Note:* You are not given the students' actual data. Your job is to develop a plan. This task requires careful thinking, and the following process is suggested:

   a. First, brainstorm different hypotheses to determine relative steepness.
   b. Second, discuss the pros and cons of those hypotheses that one or more members of the group feel are worth pursuing.
   c. Last, select one hypothesis and discuss a method for determining its validity. For example, let's say you pursue method *x* for determining which of two slopes is steeper. What do you need to do in order to determine whether this method is valid?

2. Meet with another group to compare hypotheses. If your hypotheses are identical, discuss your method for assessing the hypothesis. If your hypotheses are different, debate the two hypotheses until you come to a conclusion—one group decides that its hypothesis is not valid, each group decides that its hypothesis is or is not valid, or you reach an impasse.

3. Consider the data from two ramps:

Ramp	Height	Base	Ramp
A	8 cm	20 cm	21.5 cm
B	12 cm	26 cm	28.6 cm

   Use your present method to predict which ramp is steeper. If another group has a different hypothesis, debate with that group to determine whether both hypotheses are valid, both are invalid, or one is valid and the other invalid.

4. After the discussion with other groups, do you wish to revise your hypothesis? If so, what convinced you that your previous hypothesis was not valid?

5.   Your instructor will select one or more hypotheses for you to critique (that is, to examine carefully). For each hypothesis, decide whether you believe it is or is not valid. If you think it is valid, explain why. If you think it is not valid, explain why not or give a counterexample. If you think the hypothesis is invalid but has merit, try to modify it so that it is valid. Explain why you think the modified hypothesis is now valid.

6.   Now the class has a new problem. Let's say most of the groups doing the experiment recorded their data in inches, but some groups used centimeters. Can we determine the relative steepness with their data, or do we first have to convert one set of data into the other unit—for example, centimeters to inches? Why or why not?

SECTION **6.2** **EXPLORING PERCENTS**

Open the newspaper on any given day and you will probably find percents used in more than one story on the front page. Percents are a powerful tool that enables us to compare amounts and to describe change.

EXPLORATION 6.6 **Percents**

Sales

John sees that the local department store is having a sale. He goes to the store and finds that all televisions are 25% off.

1. Describe in words what that means.
2. Let's say he is interested in a particular television that normally sells for $400. If it is 25% off, how much will he pay for it?
3. There are two common ways in which students solve this problem.

   - Ann: 25% of $400 is $100. $400 − $100 = $300. John pays $300.

   - Bela: 75% of $400 is $300. John pays $300.

   Ann doesn't understand what Bela did. How can you help her? You may, if you wish, use a grid like the one at the right.

4. Let's say that Joe gets a 5% raise and that he presently makes $8.00 per hour. One way to determine his new wage is to find 5% of 8.00 and then add that to 8.00. Using the ideas generated above, can you figure out how to determine his new wage with only one calculation? Describe the method.

Percent Decrease and Increase

Joshua is confused. He works for the Adamson Printing Company. Last year the economy was in such bad shape that all employees agreed to take a 20% cut in pay. However, this year the economy had improved so much that the company agreed to give everyone a 20% raise. Before the pay cut, Joshua was making $30,000 a year.

5. Explain why the 20% raise does not "undo" the 20% pay cut.
6. What raise would undo the 20% pay cut?
7. Determine a general formula that will tell you what percent increase will undo an x percent decrease.

Changes in Rates

In New Hampshire, where I live, there was a tremendous amount of controversy over the Seabrook nuclear power plant. The construction of the plant was held up many times. Eventually the company that made the reactor went bankrupt. It was bought out by another utility company. Part of the deal made to ensure that the new company would make a profit was that it could increase rates by *at least* 5.5% each year for 7 years.

8. Bill and Betty Olsen figured that their average monthly electric bill last year was $83.21.

    a. If they use, on average, the same amount of electricity over the next 7 years, and their bill increases exactly 5.5% each year, what can they expect their monthly utility bill to be at the end of 7 years?

    b. Meet with your partner(s) and discuss answers and solutions. If you think you would change your method in order to do a similar problem, describe how and why the new method works.

    c. Jarrad used the following method:

    $83.21 \times 0.055 = 4.57655$

    $83.21 + 4.58 = 87.79$

    $87.79 \times 0.055 = $ etc.

    Do you think Jarrad's method is valid? Why or why not?
    If not, what suggestions would you give to Jarrad? Justify your suggestions.

9. Last year the Olsens' combined income was $42,310. If their income increases by 5.5% each year, what will their income be at the end of 7 years?

EXPLORATION 6.7   **Reducing, Enlarging, and Percents**

PART 1: A broken copy machine[1]

Most copy machines allow you either to enlarge or to reduce a copy. Some machines let you determine the exact amount of enlargement or reduction. Other machines have buttons for the changes that are most commonly made. Let's say a copy machine has buttons that will enable you to make the following changes to a copy: 10%, 50%, 100%, and 200%. For example, the 10% button means that the size of the copy will be 10% of the size of the original, whereas the 200% button will make a copy that is 200% (double) of the size of the original. The 100% button is what you push when you want a copy that is the same as your original.

1.   Let's say the buttons on the machine were as shown below.

   a.   How could you make a copy that was 25% of the size of the original?
   b.   How could you make a copy that was 3 times the size of the original?
   c.   Suppose the 100% button was broken. How could you make a copy of the original that was the same size?

2.   Suppose these were the buttons on a machine, and the 100% button was broken. How could you make a copy of the original that was the same size?

3.   Suppose these were the buttons on a machine, and the 100% button was broken.

   a.   Explain why it is now impossible to make a copy of the original that is the same size.
   b.   How close can you get to the original size?

PART 2: Making all the quilt blocks the same size

I encountered this problem while I was making the quilt patterns in Chapter 8, 9, and 10. I was able to make the patterns using a graphing software program. Most of the quilt patterns that I made are based on a grid. That is, each pattern can be achieved by making a 3 × 3, a 4 × 4, a 5 × 5, or a 6 × 6 grid. To make the quilt blocks, I first made a 6 × 6 grid on my computer. When I went to make a specific block, I opened the file containing the 6 × 6 grid and used whatever I needed. Therefore, all the little squares were always the same size.

---

[1] Text adapted from "The 100% Solution." Scott Kim © 1999. Reprinted with permission of *Discover Magazine*.

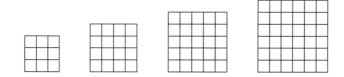

The problem I encountered is that when I wanted to print the blocks, I wanted them to all be the same size and to be smaller than the size that showed on the screen. I determined that I could get the desired size for the 6 × 6 block by telling the computer to reduce the quilt block to 25%.

1. What did I need to tell the computer so that the reduced 5 × 5 block would be the same size as the reduced 6 × 6 block?

2. What did I need to tell the computer so that the reduced 4 × 4 block would be the same size as the reduced 6 × 6 block?

3. What did I need to tell the computer so that the reduced 3 × 3 block would be the same size as the reduced 6 × 6 block?

4. I made a block starting with an 8 × 8 grid. What would I need to tell the computer so that the reduced 8 × 8 block would be the same size as the reduced 6 × 6 block?

EXPLORATION 6.8   **Mice on Two Islands**

There are two different islands in the Southern Micific Ocean, Azumi and Muremi, each of which has 20,000 mice. In this exploration, we will explore different ways in which populations can change.

1.   Brainstorm factors that could affect the rate of growth of the mouse populations.
2.   Let's say that the factors on Azumi produce an environment in which the population grows by 3000 each year, and the factors on Muremi produce an environment in which the population grows by 10% each year. That is, in Azumi's case, we can determine the next year's population additively; in Muremi's case, we can determine the next year's population multiplicatively.

     A quick computation shows that after 1 year, Azumi will have more mice—23,000, compared to Muremi's mouse population of 22,000. Do you think that Azumi will always have more mice than Muremi (disregard such factors as disease and predators)? Why or why not?

3.   **a.**   Determine a method for finding the annual mouse populations of the two islands for the next 15 years.
     b.   Determine the annual populations and compare answers and solutions with your partner(s).
     c.   If you found errors in your method, note them and explain the cause of the error. If you find that you prefer someone else's method, explain and justify the new method.

4.   Let us focus now not on the numbers of mice on each island but on the relationship between the two populations. We can then use this information to extrapolate into the future.

     a.   Let us examine two different ways to compare the relative sizes of the populations of the two islands. We could make a table showing the difference between the two populations, or we could make a table showing the ratio of the two populations. Make the two tables, and then graph the results on two different graphs. Explain each graph as though you were talking to someone who missed this exploration.
     b.   Each graph is useful for answering different kinds of questions. For example, which graph would you use to answer the following question: How much larger will Muremi's mouse population be in 5 years? Which graph would help you predict when Muremi's mouse population will be 50% greater than Azumi's mouse population? Explain your answers.
     c.   Compare your predictions with those of your partner(s). As always, if you want to change something as a result of the discussion, identify the change you wish to make and justify it.

EXPLORATION 6.9   **Do You Get What You Pay For?**[2]

There are a number of items that you buy by the pound, although you are paying for parts of the items that you don't use. For example, you throw out the shells from peanuts, the husks and cobs from corn, the rind of the watermelon, the bones from chicken, and the peels from oranges. Just what percent of the product are you tossing out?

1. Let us begin with oranges. In this step, we will only plan a procedure. Data collection takes place in Step 2.

    a. Design a procedure to find out what percent of the weight of an orange is peel.
    b. What assumptions did you make in designing your procedure? (For example, not all oranges have the same percentage of peel; like humans, some have thicker skins!)
    c. Exchange descriptions of the procedure with another person (group). Make comments on the validity of that person or group's procedure and the clarity of the description. Listen to the comments about your procedure.
    d. What changes would you make in your procedure now that you have compared it with one or more other procedures?

2. Using the procedure designed in Step 1, carry out the experiment to find the percentage of an orange that you throw away.

    a. Describe your data. Show your results.
    b. Compare your results with those of other groups or of the whole class.
    c. Is there anything that you would change about the design of your experiment? If so, explain your changes.
    d. What did you learn from this exploration?

Extensions

3. Do you think the grower makes more or less profit from thick-skinned oranges than from thin-skinned ones? Explain your reasoning.
4. Do you think grapefruits, in general, would have a higher percentage of peel than oranges? What about lemons? Explain your answer.
5. Describe how you would find the percent of waste for peeled apples.

---

[2] This exploration is adapted from one developed by Nancy Belsky.

CHAPTER 7

Uncertainty: Data and Chance

The explorations in this chapter have been designed to provide you with meaningful experiences involving the collection and interpretation of data and probability experiments so that you can grapple with some of the basic ideas in the fields of statistics and probability. I want you to experience the excitement that I have seen elementary and college students feel when they suddenly see something when graphing a set of data that they didn't see simply from looking at the numbers, or when they see patterns in chance phenomena that enable them to have a better sense of the probability of that event happening.

SECTION 7.1 — **EXPLORING, REPRESENTING, AND INTERPRETING DATA**

Statistical ideas can help you to understand data that someone else has gathered. Many people tend to accept the data they read as true. However, it is quite common for data to be inaccurate or invalid. Thus, it is important that when you look at a set of data, you think about what the data mean, who collected the data, how the data were collected, etc. We speak of the reliability of the data; that is, if two different groups were to collect the data, would they get the "same" data? We also speak of the validity of the data; that is, are the methods used to collect the data valid? In each of the following explorations, you will be presented with a set of data, and you will make and interpret graphs from those data. In each case, you will first be asked for your initial impressions of the data and for questions you have about the data; sometimes the impressions will be insights; sometimes they will be surprises. Don't be afraid to ask about reliability, validity, and numbers or sentences that you don't understand.

EXPLORATION 7.1

Population Growth, Population Density, and Area of Residence

Did you know that the framers of the U.S. Constitution required, in Article 1, Section 2, that a census be taken every 10 years? Their concept of representative government relies on knowing just who is being represented! The data these censuses collect are a window into our world.

Population Growth

1. Look at the data at the right and write down your initial impressions. That is, write at least one sentence and not more than a paragraph describing what you "see" from these numbers. Then compare your impressions with those of your partner(s).

2. Let us focus first on the total population numbers.

    a. If you were to describe the growth of the U.S. population over the last 50 years, what would you say? That is, can you be more precise than simply saying, "The population is increasing"? For example, if you say, "It is increasing fast," what do you mean by "fast"?

    b. Predict the population in the year 2000 based on these numbers. Briefly explain how you came up with your number.

3. Now let us explore how a graph helps us to see a set of data in a different light. Then we will return to the two questions just asked.

    a. Select and make a graphical representation for the population data from 1790 to 1990. Describe any problems you had in constructing the graph and how you solved these problems—for example, choosing a scale that would enable you to fit numbers as low as 3 million and as high as 250 million.

    b. Does the graph help you to see the population growth in a different light? If so, briefly explain.

    c. Compare and discuss graphs with your partner(s). If you wish to make any changes in your graph as a result of this discussion, note the changes and the reason for making the changes.

    d. Now predict the population in the year 2000 based on your impressions of the graph. Briefly describe how you obtained the number. Compare this number to your prediction in Step 2(b). Compare numbers with your partner(s). If you wish to change your prediction, note the change and the reasoning behind the change.

U.S. POPULATION, POPULATION DENSITY, AND AREA OF RESIDENCE, 1790–1990

Year	Total population	Percent increase	Pop. per sq. mi.	Percent urban	Percent rural
1790	3,929,214	N.A.	4.5	5.1%	94.9%
1800	5,308,483	35.1%	6.1	6.1	93.9
1810	7,239,881	36.4	4.3	7.3	92.7
1820	9,638,453	33.1	5.5	7.2	92.8
1830	12,866,020	33.5	7.4	8.8	91.2
1840	17,069,453	32.7	9.8	10.8	89.2
1850	23,191,876	35.9	7.9	15.3	84.7
1860	31,443,321	35.6	10.6	19.8	80.2
1870	39,818,449	26.6	13.4	25.7	74.3
1880	50,155,783	26.0	16.9	28.2	71.8
1890	62,947,714	25.5	21.2	35.1	64.9
1900	75,994,575	20.7	25.6	39.6	60.4
1910	91,972,266	21.0	31.0	45.6	54.4
1920	105,710,620	14.9	35.6	51.2	48.8
1930	122,775,046	16.1	41.2	56.1	43.9
1940	131,669,275	7.2	44.2	56.5	43.5
1950	150,697,361	14.5	50.7	64.0	36.0
1960	179,323,175	18.5	50.6	69.9	30.1
1970	203,302,031	13.4	57.4	73.5	26.5
1980	226,545,805	11.4	64.0	73.7	26.3
1990	248,709,873	9.8	70.2[1]	72.8[2]	27.2[2]

1. Estimated.   2. Figures are for 1989.

Source: John W. Wright (ed.), *The Universal Almanac* (Kansas City: Andrews & McNeel, 1992), p. 200.

Source: Excerpt from *The Universal Almanac* © 1992 by John W. Wright. Reprinted with permission of Andrews McMeel Publishing. All rights reserved.

4.   Now let us examine how graphs can enrich our understanding of the growth in population.

    a.   Make two line graphs, the first graph showing the actual increase in population from decade to decade and the second graph showing the percent increase in population from decade to decade.

    b.   The first of these graphs shows population growth from an additive perspective, the second from a multiplicative perspective. Describe the different impressions that each graph gives. Summarize the advantages and disadvantages of displaying the population increase in each of these ways.

    c.   Describe the growth of the U.S. population over the last 50 years. Then compare your response to the one you gave in Step 2(a).

Population Density

5.   **a.**   Focus now on the "Pop. per sq. mi." column. Describe your first impressions and note any questions you have about the data.

    b.   Suppose someone came up to you and said, "I don't see the point of these numbers. What do they tell you, anyway?" How would you respond to that person?

    c.   Did you learn anything from seeing different graphs or hearing others' impressions? If so, describe what you learned.

    d.   How do you think the Census Bureau came up with the number 70.2 in the initial set of data? In other words, what mathematics do you think it used?

Area of Residence

6.   Now look at the "Percent urban" and "Percent rural" columns from the initial table.

    a.   Describe your first impressions and note any questions you have about the data.

    b.   Select and make a graph that gives a sense of the increasing urbanization of the United States.

    c.   Compare and discuss graphs with your partner(s). If you wish to make any changes in your graph as a result of this discussion, note the changes and the reason for making the changes.

EXPLORATION 7.2   **Population Change**

PART 1: Changes in who we are

1. Look at the "Ancestry of U.S. Population by Rank" table.

   a. Describe your first impressions and note any questions you have about the data.

   b. Brainstorm the kinds of graphs that might illustrate these data.

   c. If you were to make a graph to display (some of) these numbers visually, describe the graph you would make and justify your choice. This at least involves making a case for the advantages of that graph and may involve explaining why you chose it over other graphs.

Ancestry of U.S. Population by Rank, 1990 Census

1990 Rank	Ancestry group	Number	Percent	1990 Rank	Ancestry group	Number	Percent
	Total population	248,709,873	100.0	33	Japanese	1,004,645	0.4
1	German	57,947,873	23.3	34	Austrian	864,783	0.3
2	Irish	38,735,539	15.6	35	Cuban	859,739	0.3
3	English	32,651,788	13.1	36	Korean	836,987	0.3
4	Afro American	23,777,098	9.6	37	Lithuanian	811,865	0.3
5	Italian	14,664,550	5.9	38	Ukrainian	740,803	0.3
6	American	12,395,999	5.0	39	Scandinavian	678,880	0.3
7	Mexican	11,586,983	4.7	40	Acadian/Cajun	668,271	0.3
8	French	10,320,935	4.1	41	Finnish	658,870	0.3
9	Polish	9,366,106	3.8	42	United States	643,561	0.3
10	American Indian	8,708,220	3.5	43	Asian Indian	570,322	0.2
11	Dutch	6,227,089	2.5	44	Canadian	549,990	0.2
12	Scotch-Irish	5,617,773	2.3	45	Croatian	544,270	0.2
13	Scottish	5,393,581	2.2	46	Vietnamese	535,825	0.2
14	Swedish	4,680,863	1.9	47	Dominican	505,690	0.2
15	Norwegian	3,869,395	1.6	48	Salvadoran	499,153	0.2
16	Russian	2,952,987	1.2	49	European	466,718	0.2
17	French Canadian	2,167,127	0.9	50	Jamaican	435,024	0.2
18	Welsh	2,033,893	0.8	51	Lebanese	394,180	0.2
19	Spanish	2,024,004	0.8	52	Belgian	380,498	0.2
20	Puerto Rican	1,955,323	0.8	53	Romanian	365,544	0.1
21	Slovak	1,882,897	0.8	54	Spaniard	360,935	0.1
22	White	1,799,711	0.7	55	Colombian	351,717	0.1
23	Danish	1,634,669	0.7	56	Czechoslovakian	315,285	0.1
24	Hungarian	1,582,302	0.6	57	Armenian	308,096	0.1
25	Chinese	1,505,245	0.6	58	Pennsylvania German	305,841	0.1
26	Filipino	1,450,512	0.6	59	Haitian	289,521	0.1
27	Czech	1,296,411	0.5	60	Yugoslavian	257,994	0.1
28	Portuguese	1,153,351	0.5	61	Hawaiian	256,081	0.1
29	British	1,119,154	0.4	62	African	245,845	0.1
30	Hispanic	1,113,259	0.4	63	Guatemalan	241,559	0.1
31	Greek	1,110,373	0.4	64	Iranian	235,521	0.1
32	Swiss	1,045,495	0.4	65	Ecuadorian	197,374	0.1

Source: *Information Please Almanac* (Boston: Houghton Mifflin, 1994), p. 836. "Ancestry of U.S. Population by Rank, 1990 Census" from *Information Please Almanac*. Source of data: Department of Commerce, U.S. Bureau of the Census. Reprinted from *The 1994 Information Please Almanac*, © 1993 by INSO Corporation.

2. Look at the "Immigrants to U.S. by Country of Origin" table of data.

   a. Describe your first impressions and note any questions you have about the data.

   b. Let us focus on a specific subset of the data: how the relative proportions of immigrants from 1951 to 1960 are similar to and different from the relative

Immigrants to U.S. by Country of Origin

(Figures are totals, not annual averages, and were tabulated as follows: 1820–67, alien passengers arrived; 1868–91 and 1895–97, immigrant aliens arrived; 1892–94 and 1898 to present, immigrant aliens admitted. 1989 and 1990 totals include legalized immigrants. (Data before 1906 relate to country whence alien came; 1906–80, to country of last permanent residence; 1981 to present data based on country of birth.)

Countries	1992	1820–1992	1981–90	1971–80	1961–70	1951–60	1941–50	1820–1940
Europe: Albania[1]	682	3,914	479	329	98	59	85	2,040
Austria[2]	701	2,662,608	4,636	9,478	20,621	67,106	24,860	2,534,617
Belgium	780	210,501	5,706	5,329	9,192	18,575	12,189	158,205
Bulgaria[3]	1,049	72,156	2,342	1,188	619	104	375	65,856
Czechoslovakia[1]	1,181	152,411	11,500	6,023	3,273	918	8,347	120,013
Denmark	764	371,787	5,380	4,439	9,201	10,984	5,393	335,025
Estonia[1]	194	1,511	137	91	163	185	212	506
Finland[1]	525	38,222	3,265	2,868	4,192	4,925	2,503	19,593
France	3,288	784,096	23,124	25,069	45,237	51,121	38,809	594,998
Germany[2]	9,888	7,078,012	70,111	74,414	190,796	477,765	226,578	6,021,951
Great Britain	19,973	5,135,957	142,123	137,374	213,822	202,824	139,306	4,266,561
Greece	1,858	698,594	29,130	92,369	85,969	47,608	8,973	430,608
Hungary[2]	1,304	1,673,817	9,764	6,550	5,401	36,637	3,469	1,609,158
Ireland	12,226	4,742,999	32,823	11,490	32,966	48,362	14,789	4,580,557
Italy	2,592	5,343,689	32,894	129,368	214,111	185,491	57,661	4,719,223
Latvia[1]	419	3,486	359	207	510	352	361	1,192
Lithuania[1]	353	4,928	482	248	562	242	683	2,201
Luxembourg[1]	25	3,214	234	307	556	684	820	565
Netherlands	1,586	376,821	11,958	10,492	30,606	52,277	14,860	253,759
Norway[4]	665	754,607	3,901	3,941	15,484	22,935	10,100	697,095
Poland[5]	25,504	665,177	97,390	37,234	53,539	9,985	7,571	414,755
Portugal	2,748	508,122	40,020	101,710	76,065	19,588	7,423	256,044
Romania[6]	6,500	227,543	39,963	12,393	2,531	1,039	1,076	156,945
Spain	1,631	283,893	15,698	39,141	44,659	7,894	2,898	170,123
Sweden[4]	1,463	1,393,971	10,211	6,531	17,116	21,697	10,665	1,325,208
Switzerland	1,023	359,385	7,076	8,235	18,453	17,675	10,547	295,680
Former U.S.S.R.[7]	43,614	3,570,704	84,081	38,961	2,465	671	571	3,343,361
Former Yugoslavia[3]	2,604	142,008	19,182	30,540	20,381	8,225	1,576	56,787
Other Europe	252	61,383	2,661	4,049	4,904	9,799	3,447	36,060
Total Europe	145,392	37,325,766	705,630	800,368	1,123,492	1,325,727	621,147	32,468,776
Asia: China[8]	38,907	1,028,247	388,686	124,326	34,764	9,657	16,709	382,173
India	36,755	548,690	261,841	164,134	27,189	1,973	1,761	9,873
Israel	5,104	138,905	36,353	37,713	29,602	25,476	476	—
Japan[9]	11,028	474,484	43,248	49,775	39,988	46,250	1,555	277,591
Turkey	2,488	414,953	20,843	13,399	10,142	3,519	798	361,236
Other Asia	262,673	4,208,748	2,042,025	1,198,831	285,957	66,374	15,729	44,053
Total Asia[10]	356,955	6,813,937	2,066,455	1,588,178	427,642	153,249	37,028	1,074,926
America: Canada and Newfoundland[11]	15,205	4,286,560	119,204	169,939	413,310	377,952	171,718	3,005,728
Central America	57,558	979,044	458,753	134,640	101,330	44,751	21,665	49,154
Mexico[12]	213,802	5,046,105	1,653,250	640,294	453,937	299,811	60,589	778,255
South America	55,308	1,379,675	455,977	295,741	257,954	91,628	21,831	121,302
West Indies	94,806	2,955,159	892,392	741,126	470,213	123,091	49,725	446,971
Other America[12]	2,676	117,078	1,352	995	19,630	59,711	29,276	56
Total America	439,355	14,763,521	3,580,928	1,982,735	1,716,374	996,944	354,804	4,401,466
Africa	27,086	412,729	192,212	80,779	28,954	14,092	7,367	26,060
Australia and New Zealand	3,205	149,303	20,169	23,788	19,562	11,506	13,805	54,437
Pacific Islands[13]	89	57,664	21,041	17,454	5,560	1,470	746	11,089
Countries not specified[14]	1,895	272,238	196	12	93	12,491	142	253,689
Total all countries	**973,977**	**59,795,158**	**7,338,062**	**4,493,314**	**3,321,677**	**2,515,479**	**1,035,039**	**38,290,443**

1. Countries established since the beginning of World War I are included with countries to which they belonged. 2. Data for Austria-Hungary not reported until 1861. Austria and Hungary reported separately after 1905, Austria included with Germany 1938–45. 3. Bulgaria, Serbia, Montenegro first reported in 1899. Bulgaria reported separately since 1920. In 1920, separate enumeration for Kingdom of Serbs, Croats, Slovenes; since 1922, recorded as Yugoslavia. 4. Norway included with Sweden 1820–68. 5. Included with Austria-Hungary, Germany, and Russia 1899–1919. 6. No record of immigration until 1880. 7. From 1931–63, the U.S.S.R. was broken down into European U.S.S.R. and Asian U.S.S.R. Since 1964, total U.S.S.R. has been reported in Europe. 8. Beginning in 1957, China includes Taiwan. 9. No record of immigration until 1861. 10. From 1934, Asia included Philippines; before 1934, recorded in separate tables as insular travel. 11. Includes all British North American possessions, 1820–98. 12. No record of immigration, 1886-93. 13. Included with "Countries not specified" prior to 1925. 14. Includes 32,897 persons returning in 1906 to their homes in U.S. *Source of data*: Department of Justice, Immigration and Naturalization Service. NOTE: Data are latest available. Source: *Information Please Almanac* (Boston: Houghton Mifflin, 1994), pp.28-29.

Source: "Immigrants to U.S. by Country of Origin" from *The 1994 Information Please Almanac* © 1993 by INSO Corporation.

proportions of immigrants from 1981 to 1990. Even now there is too much data for a clear graph, so we will focus on the following categories: Europe, Asia, Canada and Newfoundland, Latin America, Africa, and "Other." Obtain the numbers for these six categories now.

c. What can you say about changes in immigration just from seeing these numbers? Imagine that you were writing an article about changes in immigration patterns for a newspaper. What would you write?

d. Now divide your group in half. One half of the group will make a bar graph for these data, and the other half will make a circle graph. After making the graphs, present them to the group and describe what the graphs tell us that the numbers alone do not.

e. If you were to select one of the graphs to include in the newspaper article, which would it be? Justify your choice.

PART 2: Changes in how old we are

Look at the data below.

Population Distribution by Age, Race, Nativity and Sex

							Race and Nativity			
								White			
		Age									
Year	Total	Under 5	5–19	20–44	45–64	65 and over	Total	Native born	Foreign born	Black	Other races[2]
PERCENT DISTRIBUTION											
1860	100.0	15.4	35.8	35.7	10.4	2.7	85.6	72.6	13.0	14.1	0.3
1870	100.0	14.3	35.4	35.4	11.9	3.0	87.1	72.9	14.2	12.7	0.2
1880	100.0	13.8	34.3	35.9	12.6	3.4	86.5	73.4	13.1	13.1	0.3
1890	100.0	12.2	33.9	36.9	13.1	3.9	87.5	73.0	14.5	11.9	0.3
1900	100.0	12.1	32.3	37.7	13.7	4.1	87.9	74.5	13.4	11.6	0.5
1910	100.0	11.6	30.4	39.0	14.6	4.3	88.9	74.4	14.5	10.7	0.4
1920	100.0	10.9	29.8	38.4	16.1	4.7	89.7	76.7	13.0	9.9	0.4
1930	100.0	9.3	29.5	38.3	17.4	5.4	89.8	78.4	11.4	9.7	0.5
1940	100.0	8.0	26.4	38.9	19.8	6.8	89.8	81.1	8.7	9.8	0.4
1950	100.0	10.7	23.2	37.6	20.3	8.1	89.5	82.8	6.7	10.0	0.5
1960	100.0	11.3	27.1	32.2	20.1	9.2	88.6	83.4	5.2	10.5	0.9
1970	100.0	8.4	29.5	31.7	20.6	9.8	87.6	83.4	4.3	11.1	1.4
1980	100.0	7.2	24.8	37.1	19.6	11.3	83.1	n.a.	n.a.	11.7	5.2
1990	100.0	7.6	21.3	40.1	18.6	12.5	83.9	n.a.	n.a.	12.3	3.8

Source: *Information Please* (Boston: Houghton Mifflin, 1994), p. 829. "Population Distribution by Age, Race, Nativity, and Sex" from *Information Please Almanac*. Source of data: Department of Commerce, U.S. Bureau of the Census. Reprinted from *The 1994 Information Please Almanac*, © 1993 by INSO Corporation.

1. Describe your first impressions and note any questions you have about the data.

2. a. Now let us focus on understanding the phenomenon that is called "the aging of the United States." First, describe what you know about this phenomenon and why it is a concern.

    b. Brainstorm the kinds of graphs that might illustrate these data — that is, illustrate the changes in the proportions of the population that are of different ages: under 5, 5–19, 20–44, 45–64, and 65 and over.

    c. Let us now focus on three years: 1900, 1950, and 1990. Select a way to represent these data graphically. Then make the graph(s) and justify your choice.

    d. Compare your graph(s) with those of other students. Note any insights that arise from the resulting discussion.

e. Write a short paragraph describing the aging of the population that would accompany your graph from Step 2(d).

Looking Back

1. Describe the most important mathematical insights you had as a result of doing Explorations 7.1 and 7.2.
2. Describe one or two things that you learned about the United States from these explorations.
3. Did these explorations make you curious about other aspects of the U.S. population? Go to an almanac or abstract to find data that might answer your questions about these other aspects of population. Select one group of data and present those data and one graphical display of the data. This report needs to include the following:

   a. The data and your first impressions of the data.
   b. Questions you have about the data, concerning reliability, validity, or numbers or sentences that you don't understand.
   c. Your graph, your justification of your choice of that graph, and a verbal summary of the graph, as if in a newspaper article accompanying the graph.
   d. The two most important things you learned from examining the data and making the graph.

SECTION 7.2
EXPLORING DISTRIBUTIONS: CENTERS AND SPREADS

We encounter the concept of "average" frequently in our daily lives. We may think of *average* in nonmathematical ways, as meaning "so-so," or "typical." In statistics, *average* has a specific meaning beyond simply "adding all the numbers and dividing."

The following exploration will make the topic of statistics and related concepts feel more real to you. The NCTM standards say that "Instruction in statistics should focus on the active involvement of students in the entire process: formulating key questions; collecting and organizing data; representing the data using graphs, tables, frequency distributions, and summary statistics; analyzing the data; making conjectures; and communicating information in a convincing way" (*Curriculum Standards*, p. 105).

EXPLORATION 7.3   **Typical Person**

The word *typical* conjures up many of the same thoughts that the word *average* does. In this exploration, you will develop a profile of the typical person in your class by analyzing data that you and your classmates decide to collect.

1. Briefly describe topics about which you would like to collect data—for example, number of siblings, place of birth, number of schools attended in grades K–12, favorite TV show, and so on.
2. Having selected a topic, come up with the question you will ask your classmates. You want to make sure that the question is clear and that everyone will interpret the question in the same way. For example, if your topic is "number of brothers and sisters," how might others interpret this statement in different ways?
3. Analyze your data.

   a. First, discuss what kind of graph you want to make. Make the graph and briefly describe why you chose this graph over other graphs.
   b. Determine the mean, median, and mode, as appropriate. How did you determine them? Does one seem to be a more appropriate measure of the typical student than the others? Explain.
   c. Describe any challenges you had with the question, with the data, or with making the graph. That is, describe what the challenge was and how you met the challenge.
4. Prepare a presentation for the class. Include your question, your results from Step 3, and the words you will say in your presentation.

Looking Back

1. *Looking back on your presentation*

   a. Describe the strengths of your group presentation.
   b. Decide whether any of the following changes would make your presentation better: improvements in your graph, a different graph, changes in your presentation of your graph, or your determination and/or interpretation of mean, median, and mode.

2. *Looking back at other presentations*

   a. Select the group presentation that you thought was the best. Describe what was so good about that presentation.
   b. Select a presentation in which the students chose a graph that you did not think was the best choice for those data. Describe the topic and graph.

Describe the kind of graph you think would be more appropriate, and explain why.

3. *Looking back on Exploration 7.3*

    a. Describe the most important thing you learned about mathematics from this exploration and how it came about.
    b. Describe one way that you look at collecting, analyzing, and/or presenting data differently as a result of this exploration.
    c. Which data about characteristics of the typical student surprised you the most? Explain.

Extension

If you were to write a story for the campus newspaper about the typical elementary education major on your campus, how might you collect the data?

EXPLORATION 7.4   **Data Collection**

1. Select and refine a question for which collection of data will provide some "answers." For example, "How much sleep do I get?" can be refined to "What time do I go to bed and what time do I wake up?"

2. Decide how to collect and represent your data.

   a. What data will you collect?

   b. How will you collect and record the data?

   c. How precise will your "measurements" be? It will help at this time to look back and see whether the data you are collecting and how you are collecting the data will enable you to answer the question(s) you are asking.

   d. Think about how you will represent the data. For example, how will you graph them and how will you show measures of center and variation? (*Note:* You don't have to analyze all the data you collect. A student choosing to collect data on her sleep might sum up Step 2 with "I will make a chart that will have the day, my bedtime, and my waking time. I will round the times to the nearest five minutes."

3. Collect the data.

4. Analyze and represent the data. See the suggestions in Step 2(d).

5. Your instructor will specify the format for your report.

EXPLORATION 7.5   **Tailgating**

Most likely, either your parents or your driving instructor warned you about the dangers of tailgating. How do you know whether or not you are too close to the car in front of you? How close is too close? We will use data to address these questions in this exploration.

1.   Describe with your partner(s) the guidelines concerning tailgating that you presently use. Do those guidelines make sense? If they do not make sense to you, what is wrong with them?

2.   Let's say we wanted to gather information to determine a safe distance between two moving cars. Discuss the problem with your partner(s).

    a.   Define the question. This includes complexities (for example, we need more distance at higher speeds).

    b.   Determine the information you will need to obtain (through research) to answer the question.

    c.   Discuss your ideas with another group. Note any ideas that you got from the group discussion.

3.   Let us focus on several key aspects of this question. Discuss how you would measure reaction time. Meet with another group and share ideas.

    a.   Design a method for determining reaction time.

    b.   Gather the data. Discuss and address issues of precision.

    c.   Analyze the data.

    d.   Develop and justify your conclusions about reaction time.

    e.   Describe any questions you still have about determining and communicating how far is a safe distance.

    f.   Share your conclusions and questions with other groups.

4.   Describe how you can translate your data into recommendations that people can use. For example, if you say that there should be x feet between you and the car ahead of you at 60 miles per hour, you need a way to describe how a person can "measure" x feet while inside the car!

5.   Determine your recommended distance between cars at 20 mph, 30 mph, 40 mph, 50 mph, and 60 mph. Are the speeds and recommended distances proportional? How do you know? Should they be?

Looking Back on Exploration 7.5

Describe the most important thing you learned from this exploration.

EXPLORATION 7.6 Designing and Conducting a Survey

In this exploration, you will design, conduct, and present results of a survey. My students report that this is both an interesting and a challenging project. In the first edition of the textbook, I did not have this exploration, because I knew that within the time constraints of this course, it is virtually impossible to determine a truly random sample. However, when I tried surveys with my class, I found that one of the major learnings from the exploration was an appreciation of the complexities of doing a survey and a more critical eye when reading and listening to the results of surveys. Because we are presented with results of surveys daily, in newspapers and television, and because doing surveys is becoming increasingly common in elementary school textbooks, I decided to include a survey exploration in this edition.

1. Select a theme and at least four questions related to that theme. Carefully consider the questions that you will ask. You need to make sure that the respondents are answering the question you think you are asking! For example, when you ask whether they like the college, people define *like* in different ways. Similarly, you need to consider whether you want the question to be open-ended or forced choice. For example, one or more groups in my class always collect data about alcohol use. If you ask how much alcohol someone has consumed in the past week, and some respondents write, "more than 10 drinks," this presents problems when you try to determine the average.

   You also need to determine whether you will ask respondents to fill out a questionaire or to respond verbally. *Note*: when you ask questions that are sensitive, such as questions about alcohol consumption, you are more likely to get honest responses if the respondents feel some sense of anonymity. For example, you can have them fill out the survey, fold the paper, and place it into a box with other surveys.

   Essentially, you want to minimize the many problems that may occur in the data, (such as missing data when someone completes only some questions and "bad" data when someone gives a response that can't be used or quantified).

2. Determine your target population. In most cases, your target population will be students at your college.

3. Devise, describe, and justify a strategy to get a representative sample. This discussion involves both logistics and justifications.

   a. Logistics: Where, when, and how will you collect the data?
   b. Justification: Why do you think the where, when, and how will give you a representative sample?

4. Determine how you and your group members will code the data. For example, if you want to compare males and females, then you have to code their responses separately.

5. Collect the data.

6. Analyze the data.

   a. First, analyze the data from your own sample. Determine the centers and spreads for each question, and make sketches of appropriate graphs for each question.
   b. Compare results with other members of your group. Are your results (such as centers and spreads) close or not?

7. Combine the data that you and your group members have collected. Determine what you have found and how best to present your findings to the class: centers and spreads, graphs, the text of your presentation, and the like.

8. Present your report to the class. Your instructor will specify the format for your report.

SECTION ◆ **7.3** ◆   **EXPLORING CONCEPTS RELATED TO CHANCE**

Chances are, you have some understanding of probability concepts. For instance, you know that when you roll dice, the probability of getting a 2 is less than the probability of getting a 7. You know that the probability of snow is less in Atlanta than in Buffalo.

Like the concept of "average," "chance" also has a specific meaning that you will explore in this section. In the process, you will use some tools you have developed in earlier chapters and develop some new ones.

EXPLORATION 7.7   **Using Sampling to Estimate a Whole Population**

Scientists have used the idea of sampling and proportions to estimate various wildlife populations. This exploration is a simulation of a method that has been used by naturalists to estimate the number of fish in a lake.

Your group will obtain a supply of "fish" from your instructor, and you will estimate the total number of fish without counting every single one.

1.   Obtain your "fish" from your instructor and place them in a container. The container represents the lake in which these fish are distributed. Imagine that there were concerns that the lake's fish population was declining and that this would hurt tourism, which would in turn hurt the local economy. You have been asked to determine the number of fish in the lake. This number will be used as a baseline, so that, when the procedure is repeated in succeeding years, analysts will be able to see whether the population is actually declining.

   a.   Take a few minutes to brainstorm possible ways to estimate the number of fish in the lake by a means other than counting all of them.
   b.   Select one of your group's ideas and try it out on the fish your instructor gave you. Briefly write up your group's method, your justification of the method, and an estimate of the number of fish you got when you applied your method.

2.   Now we will explore a method that scientists have used. First, they catch a "reasonable" number of fish and then tag and release them. Then they catch another batch of fish; some of these fish will have tags and some will not. We now have enough data to calculate an estimate of the number of fish in the lake.

   a.   Discuss this procedure in your group until you understand both how and why it works. You may want to do a trial run (using the problem-solving strategy "act it out") to help you better understand the procedure.
   b.   How do you think the scientists determine the "reasonable" number of fish to tag?

3.   Now use the procedure with the fish that your instructor gave you.

   a.   Decide how many fish to tag and briefly justify your choice. Tag that number of fish.
   b.   Determine the size of your sample — that is, how many fish to catch — and briefly justify your choice. Catch the fish and record the number of caught fish that are tagged and the number that are untagged.
   c.   Use this information to estimate the total number of fish in the lake.

d. Repeat parts (b) and (c) until you feel that your estimate is "close." Briefly justify why you think your estimate is close. You may want to use a table like the following:

Sample	Number caught	Number caught that were tagged	Estimate of actual population
1			
2			
3			

4. Have a class discussion in which each group presents its estimate, justifies its choice of numbers of fish to tag and to catch again, and explains its degree of confidence that its estimate is close to the actual population. What did you learn from the class discussion?

5. Imagine doing this in a real lake with real fish.

   a. Discuss how many fish it would be reasonable to catch and tag. Describe your group's decision and the rationale for that decision.

   b. In real life, what factors might cause the tagged fish not to be randomly distributed in the lake? How might the biologists deal with these issues?

EXPLORATION 7.8   **Predicting Population Characteristics from Sampling**

Your group has received a bag containing at least two kinds of objects. Your goal is to estimate, through sampling, the proportion of each kind of object. For example, if there are some red chips and some blue chips, your task is to estimate the actual proportions of red and blue chips in the whole "population" of your bag.

1.   Obtain your bag from your instructor. Discuss how large you want your sample size to be. Describe your choice and the rationale behind your choice.
2.   Obtain and record a sample. Determine your estimate of the actual proportions of the different objects in the bag at this point, or say "not enough data." Describe any insights or questions you have. Repeat this process until you feel confident that you can predict the actual proportions. Summarize your data and your reasoning. You might want to use a table like the one below.

Sample number	Sample size	Number of each object			Predicted actual proportion of each object		
		A	B	C	A	B	C
1							
2							
3							
Totals							

3.   Meet with another group that had the same objects. Compare your data and your conclusions. Describe any additional insights or questions you have as a result of your discussion.
4.   Have a class discussion in which each group presents its estimate and explains its degree of confidence that its estimate is close to the actual proportions. Describe any additional insights or questions you have as a result of the presentations and discussion.

EXPLORATION 7.9 **Buffon's Needle**

In the eighteenth century, the French naturalist Comte de Buffon (1707–1788) performed an experiment that is still interesting today.

The question that he investigated was: If a needle is dropped on a floor made of planks (floorboards), what is the probability that the needle will touch or fall across one of the cracks in the floor when the length of the needle is less than the distance between the planks ($l < d$)? He presented this problem at the University of Paris in 1760 as one way of approximating π.

1. Paraphrase Buffon's question in your own words. Then write down your ideas on how to simulate this experiment.
2. Your instructor will give you directions for re-creating this experiment with a toothpick and lined paper instead of a needle and a wood floor.
3. Before you perform the experiment, respond to the following questions.

   a. What is your estimate of the probability that the toothpick will land on or touch a line? Explain your reasoning.
   b. How many times should you drop the toothpick in order to be "relatively" close to the actual probability? Explain your reasoning.

4. Now conduct the experiment with your group. Drop the toothpicks onto the paper from a height of about one foot. If any toothpicks miss the paper, drop them on the paper again. Count and record as "hits" the number of toothpicks that touch or cross the lines.

   Do 10 trials. Each trial consists of the number of hits out of 10 drops. Record your results in a table like the one at the right.

5. What percent of the drops were hits? (This is called the experimental probability.)
6. In a moment, we will gather the data from all groups. However, before we do that, answer the following question. If we were to make a histogram from 10,000 trials, what do you think the distribution of that data would look like? Would it be uniform, normal, bimodal, of constant slope up (or down), random, or something else? Briefly describe your reasoning in picking that distribution.
7. Collect the data from all the groups in the class. What is the class's experimental probability?

Trial	Number of hits
1	
2	
3	
4	
5	
6	
7	
8	
9	
10	

8. Buffon found that when the length of the needle is half of the distance between the floorboards, the probability of its touching or crossing a crack is $1/\pi$, or about 32%.

   a. Using this information, calculate your approximation of π from the class data.
   b. Determine the percent error.

9. Suppose the needles are the same length as the distance between the planks.

   a. On what percent of the drops do you predict that the needle will hit a line? Explain the reasoning behind your prediction.

   b. Repeat the experiment with toothpicks that are the same length as the distance between the lines. Write down the results of the experiment.

10. Design an experiment in which the toothpicks and the distance between the planks have a different ratio.

   a. Describe your experiment.

   b. Describe your findings.

   c. Based on your data, can you make a general statement that will allow us to find the probability of a needle touching or crossing a line that can be applied to a needle of any length and any distance between the boards? If so, write it down both in words and as a formula.

11. Although this was a probability experiment, you used other mathematics concepts and topics. What were they? How did you use them?

12. Describe the most important thing you learned from this exploration.

EXPLORATION 7.10   **How Many Boxes Will You Probably Have to Buy?**

Cereal companies often place prizes inside boxes of cereal to attract customers. One of these promotions lends itself nicely to exploration. Many elementary school teachers have done this exploration, in a simpler form, with their students. It certainly captures their attention!

PART 1: Three prizes

Inside specially marked packages of Sugar Sugar cereal, you can get a photograph of one of three sports figures. Many kids will want to have one of each.

1. How many boxes do you think you would have to buy in order to get one of each? Brainstorm this with your partner(s) and write down your thoughts.
2. We will explore a slightly different question: If you were to buy boxes of cereal until you had all three photographs, how many boxes would you have to buy, on average? How can we figure this out? With your group, devise a plan, describe your plan, and justify your plan.
3. We can simulate this problem in many different ways. However, each of the ways requires that we use random samples. We can rephrase this restriction by saying that we want to simulate this problem by *randomizing*. Brainstorm ways to simulate this problem. Then describe and justify your plan.
4. **a.** Now that you have a plan for simulating the problem, you need to decide on a plan for recording your data. Describe and justify your plan.
    b. Meet with another group to share ideas. If this sharing inspires you to make any changes in how you will record your data, write these changes down.
5. Record your group's data and conclusions.
6. Take the data from all the groups and record those data in a chart like the one below. But first, predict the shape of the distribution of the data and explain your prediction.

Number of boxes that had to be bought	3	4	5	6	7	8	9	10	11	12	etc.
Frequency											

7. Use the data from the whole class.
    a. Make a histogram and determine the mean and standard deviation. Briefly describe what the histogram and standard deviation tell us about the data.
    b. Make a boxplot. Briefly summarize what the boxplot tells us about the data.
    c. Describe the advantages and disadvantages (or limitations) of each graph.
8. From your data, you can answer questions like: What is the probability of getting all 3 photographs after buying 3 boxes of cereal? after buying 4 boxes? and so on. As you discovered, not every group obtains the same probabilities. These probabilities that you have determined are called experimental probabilities; that is, the numbers come from your experimenting. The Law of Large Numbers states what many people conclude from their own common sense: The larger the sample size, the more likely it is that the experimental probability will be accurate. In this situation, we can also determine the theoretical probabilities. For example, the theoretical probability of getting each of the photographs is 1/3, assuming, of course, that the numbers of all of the three photographs are equal

and that the company mixes up the boxes well, so that each store is likely to have equal numbers of the three photographs.

a. Take some time now with your partner(s) to determine the theoretical probabilities of getting 3 different photographs after buying 3 boxes.

b. Describe your method and the problem-solving tools that you used. Discuss your method with another group.

c. Now determine the theoretical probabilities of getting 3 different photographs after buying 4 boxes and after buying 5 boxes. I will give an important hint: It is very important to look for patterns in your representation and to be systematic. The object of this part of the exploration is not so much to get the theoretical probabilities as it is to understand the importance of the problem-solving tools of appropriate diagrams, looking for and using patterns, and being systematic.

d. Extrapolate from your work to predict the theoretical probability after 6 boxes. Describe the reasoning that you used to make your prediction.

e. Describe what you learned about problem-solving from this step of the exploration.

9. Compare the theoretical and experimental probabilities of getting 3 different photos after buying 3 boxes, 4 boxes, and 5 boxes. Why aren't they identical? Which do you trust more? Explain your responses.

PART 2: Extensions

1. Repeat this exploration, assuming that Sugar Sugar cereal offers six different photos, rather than three.

2. Here are one group's data for 25 turns for the variation with six photos.

Number of boxes needed to get 6 photos	6	7	8	9	10	11	12	13	14	15	16	17	18	19	20	21	22	23	24	25	26	27	28	29	30
Frequency	0	2	1	2	2	4	2	3	1	1	2	1	2	0	0	0	0	1	0	0	0	0	0	0	1

a. Determine a way to find the mean without adding all 25 numbers and dividing by 25.

b. Using boxplots, determine whether 23 and/or 30 are outliers.

c. Without adding up the remaining 23 numbers, determine the new mean if we threw out the turns of 23 and 30. Explain how you did this.

Looking Back on Exploration 7.10

1. Describe the most important things you learned from this exploration.

2. Describe any questions that arose from this exploration. These could include "what-if" questions.

EXPLORATION 7.11   **More Simulations**

The following situations present real-life questions for which determining the theoretical probability is either tedious or impossible. Thus, they lend themselves to simulations in which we can determine the experimental probability.

For each of the following questions, develop and execute a simulation plan to help you answer the question. Your instructor will specify the format for your reports.

1.   *The basketball game is on the line*   Horace has just been fouled, and there is no time on the clock. His team is down by one point, and he gets two shots. Horace is an 80% free-throw shooter. What percent chance of winning in regulation time does his team have?

2.   *Genetics*   Maggie and Tony have just discovered that they are carriers of a genetically transmitted disease and that they have a 25% chance of passing on this disease to their children. They had planned to have 4 children. If they still have 4 children, what is the probability that at least one of the children will get the disease?

3.   *Overbooking*   Airlines commonly overbook; that is, they sell more tickets for a flight than there are seats.

   a.   Why do you think airlines overbook?
   b.   An airline has a plane with 40 seats. What information will help it to decide how much to overbook?
   c.   Using the information from your instructor, develop a simulation plan to help the airline decide how many tickets to sell. Meet with another group to share ideas. Note any changes in your plan.
   d.   Do the simulation. What is your conclusion? Support your conclusion.

4.   *Having children*   In an attempt to deal with overpopulation, the Chinese government has a policy that couples may have only one child. One of the biggest problems with this policy is that in Chinese culture, having a boy is preferable. As a result of the policy and the preference for boys, infanticide is not uncommon in China (and in other countries too — for example, in parts of Nepal, where I was in the Peace Corps). That is, it is not uncommon for parents to kill their newborn baby if it is a girl. Obviously there are many ways to deal with this problem.

   a.   One way would be to modify the policy to say that a couple could keep having children until they got a boy. However, one mathematical problem with this policy is that it would affect the ratio of men to women. With your partner(s), discuss how this policy would affect the ratio of boys to girls (assuming that there is a 50% chance of having a boy or a girl). Explain, as though you were talking to someone who does not understand, how this policy would affect the ratio.
   b.   Assume that this policy was implemented. Develop and run a simulation to determine the ratio of boys to girls after the implementation. That is, theoretically, the ratio of boys to girls will be 1 : 1. If this policy were implemented, what would be the ratio of boys to girls among those who were born after the implementation?

SECTION   **7.4**   **EXPLORING COUNTING AND CHANCE**

Many real-life questions or problems fall into the field of probability called combinatorics. That is, there are patterns that enable us to determine the number of elements in the sample space and the number of elements in the subset in which we are interested.

EXPLORATION 7.12   **License Plates**

Some of the readers of this book will have grandparents who can remember life without the automobile, something most of us cannot imagine.

    When cars first started being sold, there were no license plates and no driver's licenses. If you could afford a car, you could drive it! However, as cars became more popular, the need to require people to register their cars arose. For example, police could thus identify cars that were being driven unsafely.

    The evolution of license plate configurations has varied from state to state. Many elementary and middle school teachers have found the origin of license plates an extremely interesting topic to explore with their students. It turns out that different states made different decisions about what to do when they ran out of combinations. Let's explore some of the configurations that have been used.

1.   **M 728**

    a. An early license plate configuration consisted of a letter of the alphabet followed by 3 digits. With this system, how many possible license plates could a state make?

    b. A next step would be to simply add a digit—that is, use a letter followed by 4 digits. How many different license plates could be made this way?

    c. Let's say that your state uses the system in part (b) and is about to run out of license plates. What would you recommend next? Why?

2.   **123 456**

    a. At some point, most states have used the following system: License plates consisting of a 6-digit number. How many possible license plates are there in this system?

    b. Let's say your state has been using the system in part (a) and is about to run out of license plates. What would you recommend next? Why?

3.   **3 G 2346**

    a. Some states went for the pattern shown above: a digit, then a letter, followed by 4 more digits. How many possible license plates are there in this system?

    b. Many states discovered that 3 letters and 3 digits makes a sufficient number of combinations. How many possible license plates are there in this system?

4.   Even with all these combinations, some states ran out of possible combinations. What would you recommend next? Why? How many different license plates could be made with your proposed system?

5.   Let's say a state outgrew the license plate consisting of 3 letters and 3 digits and was considering these two options: 1 letter, 2 digits, and 3 letters, or 6 letters. Which would you recommend? Why?

6.   Let's say a state decided to make a license plate that consisted of 6 letters. Theoretically, this would make $26 \times 26 \times 26 \times 26 \times 26 \times 26$ possible license plates. However, Mason has a problem. His calculator doesn't have enough

spaces to show the exact answer; it displays only 8 digits. It shows 3.0891578 08. Explain how you could determine the actual answer with a *minimum* of pencil and paper computation.

7.   Let's say a state decided to have a license plate with the following format: 2 letters followed by three digits followed by 1 letter—for example, AB123C.

    a. How many different license plates could be made with this system?

    b. Maegen says that the answer here is the same as that for 3 letters followed by 3 digits. What do you think? Justify your answer.

EXPLORATION 7.13   **Native American Games**

Human beings have played games of chance for thousands (perhaps tens of thousands) of years. In 1975, Stewart Culin wrote a book called *Games of the North American Indians*.[1] I was fascinated by the many different kinds of games of chance and the similarities and differences between games played by different tribes. This exploration uses some of the games in Culin's book and will help you to grapple with basic probability concepts, have some fun, and gain a sense of history.

Each group will learn to play one of the games below and then explain the game to the rest of the class and report what they have learned about the game from playing and analyzing it. Each group will explore its game within the following parameters:

1. *Predictions*   After playing for only a few throws, predict the answers to the following questions. Once you have made your own prediction, discuss your predictions in your group.

    a. What is the probability that you will score at least one point on your turn?
    b. What is the probability that you will score no points on your turn?
    c. How many different outcomes are there?

2. *Experimental probability*   Play the game for some time and determine answers to the first two questions in Step 1. That is, determine the experimental probabilities.

3. *Theoretical probability*   Determine the theoretical probabilities for all three questions in Step 1.

4. *Presentations*   In your presentation to the rest of the class, you must:

    a. Demonstrate your game.
    b. Give your answers to the three questions in Step 1, making sure that the listener can understand how you got your answers. You are free to make tables, charts, diagrams, and whatever other visual aids you consider helpful.
    c. Share any other insights or questions about the game.

PART 1: A game from the Klamath tribe in California[2]

Four sticks are marked as shown at the right. The two sticks at the top are known as shnawedsh, (women), and the two bottom sticks are called xoxsha or hishuaksk, (men). The lines were made by pressing a hot, sharp-pointed tool against the wood. The other side of each stick is unmarked (plain).

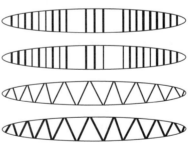

The game consists of throwing the four sticks. The scoring is as follows:

- If all four marked sides face up, 2 points.

- If all four plain sides face up, 2 points.

- If both male sticks are up and the female sticks down, 1 point.

- If both female sticks are up and the male sticks down, 1 point.

You may make your own sticks or select other materials that will create an equivalent game.

[1] Stewart Culin, *Games of the North American Indians* (New York: Dover Publications, 1975). Excerpt from *Games of the North American Indians* by Stewart Culin. Copyright © 1975. Reprinted by permission of Dover Publications, Inc.
[2] Culin, p. 136. Excerpt from *Games of the North American Indians* by Stewart Culin. Copyright © 1975. Reprinted by permission of Dover Publications, Inc.

PART 2: A game from the Nishinam tribe in California[3]

Two acorns have been split lengthwise in halves, and the outsides have been painted red or black.

The game consists of throwing the four acorn halves. The scoring is as follows:

- If all four painted sides face up, 4 points.
- If all four painted sides face down, 4 points.
- If three painted sides face up, no points.
- If two painted sides face up, 1 point.
- If one painted side faces up, no points.

In this game, a player keeps throwing until a throw results in 0 points, and then it is the next player's turn.

You may make your own acorns or select other materials that will create an equivalent game.

PART 3: A game from the Songish tribe from British Columbia[4]

Four beaver teeth have been marked in the following manner on one side:

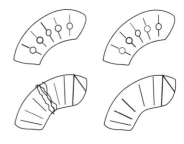

- Two teeth have been marked with a row of circles; they are called women.
- The other two teeth have been marked with cross lines; they are called men.
- One of the male teeth is tied in the middle with a small string and is called ihkakesen.

The game consists of throwing the four teeth. The scoring is as follows:

- If all four marked sides face up, 2 points.
- If all four marked sides face down, 2 points.
- If both male teeth are up and the female teeth are down, 1 point.
- If both female teeth are up and the male teeth are down, 1 point.
- If the ihkakesen is face up and the other teeth face down, 4 points.
- If the ihkakesen is face down and the other teeth face up, 4 points.

In this game, a player keeps throwing until a throw results in 0 points, and then it is the next player's turn.

You may make your own "teeth" or select other materials to create an equivalent game.

PART 4: A game from the Zuni tribe in New Mexico[5]

Four sticks have been painted red on one side; the other side is plain.

[3] Culin, pp. 154–155. Excerpt from *Games of the North American Indians* by Stewart Culin. Copyright © 1975. Reprinted by permission of Dover Publications, Inc.
[4] Culin, p. 157. Excerpt from *Games of the North American Indians* by Stewart Culin. Copyright © 1975. Reprinted by permission of Dover Publications, Inc.
[5] Culin, p. 223. Excerpt from *Games of the North American Indians* by Stewart Culin. Copyright © 1975. Reprinted by permission of Dover Publications, Inc.

The game consists of throwing the four sticks. The scoring is as follows:

- If all four painted sides face up, 4 points.
- If three painted sides face up, 3 points.
- If two painted sides face up, 2 points.
- If one painted side faces up, 1 point.

In this game, a player keeps throwing until a throw results in 0 points, and then it is the next player's turn.

You may make your own sticks or select other materials that would create an equivalent game.

PART 5: Another game from the Zuni tribe in New Mexico[6]

Three sticks have been painted red on one side and black on the other side.

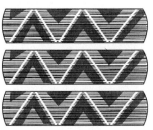

The game consists of throwing the three sticks. The scoring is as follows:

- If the three red sides face up, 10 points.
- If the three black sides face up, 5 points.
- If two red sides and one black side face up, 3 points.
- If one red side and two black sides face up, 2 points.

In this game, if three red sides face up, the thrower gets another turn.

You may make your own sticks or select other materials that would create an equivalent game.

Looking Back on Exploration 7.13

After the group presentations, each group will discuss the following questions.

1. In what ways were all of the games similar to one another—that is, what characteristics did they all have in common?
2. In what ways were some of the games different from some of the other games?
3. Describe the most important things you learned from these games.

[6]Culin, p. 221. Excerpt from *Games of the North American Indians* by Stewart Culin. Copyright © 1975. Reprinted by permission of Dover Publications, Inc.

8

Geometry as Shape

Geometry is generally underrepresented in elementary school, partly because the geometry chapters occur at the end of most textbooks and partly because more emphasis is placed on developing students' skills in computation. Another reason is that many elementary teachers have negative feelings toward this area of mathematics because of bad experiences in high school geometry. My hope is that you will be among the teachers who redress this imbalance.

At its root, geometry involves shape. In Chapters 9 and 10, we'll look at various applications of geometry and connections between geometry and other fields of mathematics. We will start with explorations that will orient you to a more geometric way of thinking.

Explorations 8.1, 8.2, and 8.3 make use of three types of manipulatives — geoboards, tangrams, and polyominoes — that are helpful in developing a strong understanding of many important mathematical ideas. These explorations have multiple parts and address topics from throughout Chapter 8. Your instructor may choose to use parts of these explorations in combination with other explorations from the individual sections, which follow these opening explorations.

EXPLORATION 8.1   Geoboard Explorations

Geoboards are a popular manipulative in elementary schools and are very versatile. Many teachers also use them as a manipulative for learning about fractions. Smaller geoboards generally contain 25 pegs (5 rows of 5 in a row), and larger geoboards generally contain 100 pegs. There are also circular geoboards, generally with 24 pegs arranged in a circle. I know many teachers who have actually had their classes make a set of geoboards to save money, and the making of the geoboards — how much material to buy, where to put the nails, and so on — is a mathematics project in itself!

In Chapters 8 through 10, we will use geoboards to explore a number of mathematical concepts and to develop other problem-solving and reasoning tools.

PART 1: Communication

This first part of the exploration both serves as an introduction to geoboards and reinforces the need for clear communication when talking about shapes.

Instructions: In this exploration, you will work in groups of 3.

Persons A and B each have a geoboard (or geoboard dot paper) and sit back to back. Person C is the observer.

Person A makes a figure on the geoboard. Next, using only words, person A gives directions so that person B can construct the same figure on his or her geoboard. The observer watches without comment and will give feedback at the end. Person B's responsibility is to ask for clarification whenever person A's directions are not clear.

1. After you are finished, compare geoboards. The figures may or may not be identical. In either case, take some time to listen to person C's feedback and to discuss strategies that might increase the likelihood of success and discuss places where communication broke down.
2. Rotate roles: Person C now makes the figure and gives directions, person A makes the copy, and person B is the observer.
3. Repeat this process once more so that each person has a turn in each role.
4. Afterwards, write down your comments on the following:
    a. Describe strategies that increased the likelihood of success.
    b. Describe something you learned from this exploration.

PART 2: Challenges

If you have played with geoboards before or if you have visited classrooms where children were working with geoboards, you may have heard someone say, "I wonder if . . ."—for example, "I wonder if you can have a figure with exactly 3 right angles." "I wonder how many different figures you can have that have only one peg inside."

The challenges below address a few of the hundreds of possible "I wonder" questions. Pursuing these challenges can help you develop problem-solving and reasoning skills and also help you to understand some of the properties of geometric figures.

In Steps 1–3, construct the figures on your geoboard (or geoboard dot paper) or explain why you think it is impossible to do so.

1. a. A figure with just 1 right angle.
    b. A figure with 2 right angles.
    c. A figure with at least 1 right angle but no sides parallel to the edges of the geoboard.
    d. A figure with 6 right angles.
2. A figure with exactly 2 congruent, adjacent sides.
3. Two figures with different shapes but the same area.
4. a. Find all possible squares that can be made on a 25 peg geoboard.
    b. Make a quadrilateral with no parallel sides.
    c. Make a parallelogram with no sides parallel to the edge of the geoboard.
    d. Make two shapes that have the same shape but are different sizes.
5. a. Make a pentagon.
    b. Make three other pentagons that have at least one more characteristic in common beyond having 5 sides.

6. **a.** Divide this region into 4 triangles. **b.** Divide this region into 3 congruent triangles. **c.** Divide this region into 3 shapes that have the same area.

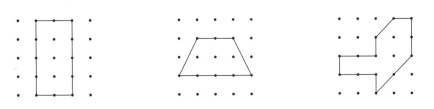

In Steps 7 and 8, construct the figures on your geoboard or explain why you think it is impossible to do so.

7. **a.** A quadrilateral that touches 5 pegs and has 1 peg in the inside.
   b. A quadrilateral that touches 5 pegs and has 2 pegs in the inside.
   c. A quadrilateral that touches 8 pegs and has 1 peg in the inside.
   d. A quadrilateral that touches 8 pegs and has 2 pegs in the inside.

8. **a.** A figure that touches 5 pegs and has 1 peg in the inside.
   b. A figure that touches 5 pegs and has 2 pegs in the inside.
   c. A figure that touches 5 pegs and has 3 pegs in the inside.
   d. A figure that touches 5 pegs and has 4 pegs in the inside.
   e. A figure that touches 5 pegs and has 5 pegs in the inside.
   f. A figure that touches 5 pegs and has 6 pegs in the inside.

9. **a.** How big can a figure be that touches 5 pegs and has x pegs in the inside?
   b. Are there combinations of touching x pegs and having y pegs in the inside that you think are impossible? If so, explain why.

10. What is the most important thing you learned from this exploration?

PART 3: Recognizing and classifying figures with geoboards

Geoboards can also be used to help students appreciate the need for names for various geometric figures and geometric ideas. If we were to look at the universe (remember Venn diagrams) of geometric shapes, there are countless possible subsets. In elementary, middle, and high school you learned the properties of a number of shapes, many of which have specific names. But a name implies certain characteristics. For example, there are all kinds of isosceles triangles, but regardless of their size and angles, all isosceles triangles have certain attributes in common. In this exploration, you explore different kinds of subsets from a small universe of geometric figures.

1. Cut out the 18 geoboard diagrams on page 213 keeping the letter that goes with each diagram.
2. Separate them into two or more subsets so that the members of each subset are alike in some way. Record the figures in each subset.
3. Describe the attributes that all the shapes in each subset have in common.
4. Name each subset. If you know of a mathematical name for a subset, use it. If you don't know of a mathematical name, make up a descriptive name that fits the subset.

5.   Your instructor will give you a new shape. Determine which subset it goes into. If there is disagreement in your group, describe it.

6.   Repeat this process as many times as you can in the time given. That is, in what other ways can we separate these figures into two or more subsets so that every subset has one or more common characteristics?

Geoboard Diagrams for   EXPLORATION 8.1,   PART 3:
Recognizing and Classifying Figures with Geoboards

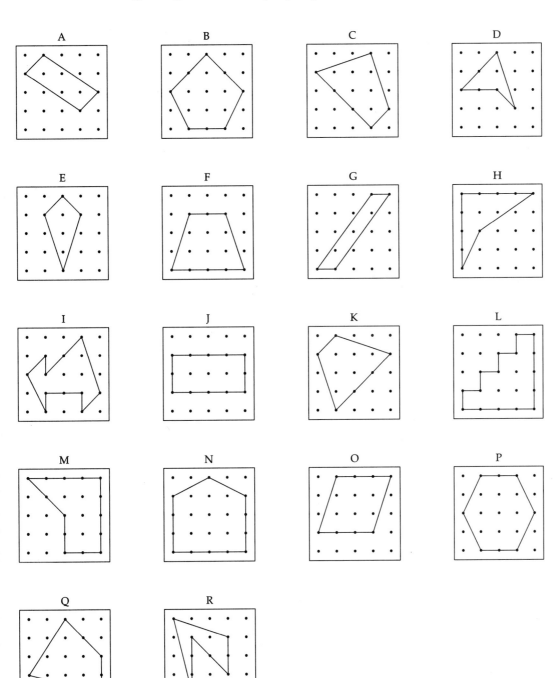

PART 4: Defining parallel and perpendicular lines

The terms *parallel* and *perpendicular* occur both in algebra (equations of lines) and in geometry (properties of many geometric figures). This exploration requires you to think about these concepts and some of the connections between algebra and geometry.

1. On each of the geoboards below, make a line parallel to the given line. Develop, describe, and justify a rule or procedure for making sure that the two lines are parallel. The rule or procedure should be more precise than "It looks parallel."

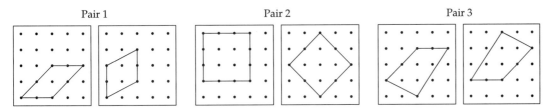

2. On each of the geoboards below, make a line perpendicular to the given line. Develop, describe, and justify a rule or procedure (beyond "it looks perpendicular") for making sure that the two lines are perpendicular.

PART 5: Congruence with geoboards

Look at the following three pairs of figures.

| Pair 1 | Pair 2 | Pair 3 |

1. Which, if any, pairs are congruent?
2. How can you tell without cutting them out and placing one figure on top of the other?
3. Discuss the question asked in step 2, and then develop a test that someone else could use to determine whether two figures are congruent.

PART 6: Quadrilaterals

Many people have negative feelings about geometry because they recall memorizing a lot of terms in elementary school and memorizing a lot of proofs in high school. This is unfortunate because the point of geometric knowledge is to have a better understanding of shapes in our everyday world, to appreciate the aesthetics in natural and human-made shapes, and (in some occupations) to apply one's understanding of shapes. This exploration takes an inductive approach: From examining many different figures, what commonalities can you see?

1.   Get out several sheets of geoboard dot paper.
      a.   Make several different squares.
      b.   Make several different rectangles.
      c.   Make several different parallelograms.
      d.   Make several different rhombi.
      e.   Make several different trapezoids.
      f.   Make several different kites.

2.   Examine each kind of quadrilateral and answer the following question: What characteristics and properties do all of the figures have in common? For example, look at sides, angles, or diagonals, to name but a few ways to look.

3.   Discuss your observations with others.

4.   Use the table on page 217 to summarize your findings and show the connections between common properties. For example, squares and rhombi both have four congruent sides, but squares and rectangles have four right angles.

Table for EXPLORATION 8.1, **PART 6: Quadrilaterals**

Characteristic	Squares	Rectangles	Parallelograms	Rhombi	Trapezoids	Kites
All 4 sides equal	X			X		

EXPLORATION 8.2   **Tangram Explorations**

Like geoboards, tangrams are very versatile manipulatives that we will use on several occasions in this and the next two chapters. Tangrams were invented in China at least two hundred years ago, but we are not sure by whom or what for. They quickly became a popular puzzle, because there are so many different things you can do with them! The English word *tangram* probably came from American sailors who referred to all things Chinese as Tang, from the Cantonese word for China.

You will use tangrams provided by your instructor or make your own set from the template at the back of the book.

PART 1: Observations, discoveries, and questions

The directions here are very simple: Play with the tangrams for a while. You may have questions for which you have some preliminary responses, and you may have some questions for which you don't have an initial response. As you explore the shapes, you will notice things: patterns and relationships among the various pieces. You might make interesting shapes. You may ask yourself various "what-if" questions.

Record your observations: patterns, discoveries, conjectures, and questions.

PART 2: Puzzles

As you may have found, you can make a number of interesting shapes with tangrams. Below are some famous puzzles that call for careful thinking.

1. **a.** Use all seven tangram pieces to make each of the five figures shown on page 221. Sketch your solution.
    b. Describe any thinking tools that you became aware of while figuring out how to make the figures.
    c. Describe any new observations: patterns, discoveries, conjectures, and questions.

2. This is a famous tangram paradox. Each of the figures shown at the bottom of page 221 has been made with one tangram set. Yet it appears that the figure at the right has one more piece.
    a. Solve the puzzle and show your solution. Describe any thinking tools that you became aware of while solving this puzzle.
    b. Describe any new observations: patterns, discoveries, conjectures, and questions.

3. Make your own puzzle. Name it. Give it to someone else to solve.
4. Make a figure with the tangram pieces. Write directions for making that figure, as though you were talking on the phone to a friend. Exchange directions with a partner. Try to make the figure from your partner's description.
    a. Discuss any problems that either of you encountered in understanding what the other had written.
    b. Now make a new figure and write directions for making that figure, as though you were talking on the phone to a friend.

Figures for EXPLORATION 8.2, **PART 2: Puzzles**

1.

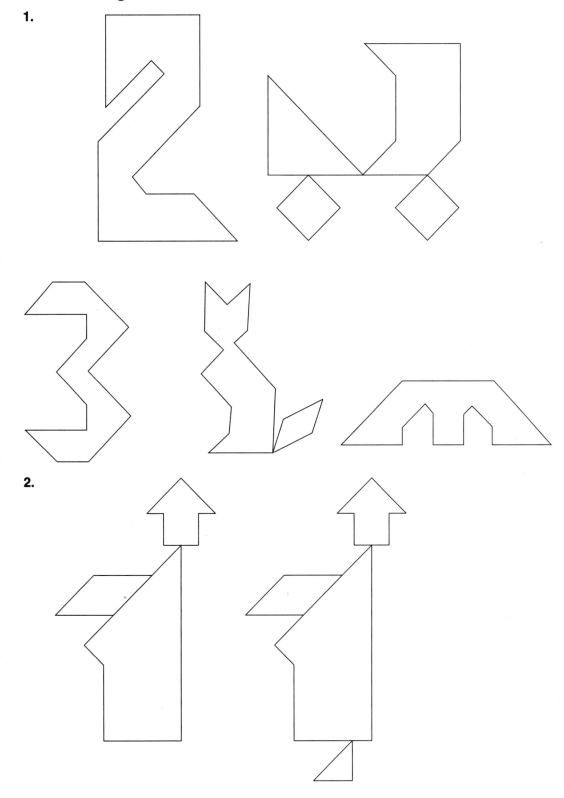

2.

PART 3: Classifying and naming geometric figures with tangrams

This exploration uses tangrams to explore names and properties of different shapes.

1. *The two small triangles and the medium triangle*   Using only the two small triangles and the medium-size triangle, make as many shapes as you can, with the following restriction: When two pieces meet, the whole side of one piece must fit against the whole side of another piece. Record your findings on a separate piece of paper by tracing the shapes. Try to be systematic to ensure that you find all possible shapes.

    a. Describe your strategies (other than random trial and error) for finding all possible shapes. Try to explain your strategies so that someone reading this description via email could understand it.
    b. Name each shape. If you know of a mathematical name, use it. If not, make up a name that is related in some way to the shape.
    c. Describe your other observations from this exploration: patterns, discoveries, conjectures, and questions.

2. *The two small triangles and the square*   With the same restrictions, make as many shapes as you can using only the two small triangles and the square. Record your findings on a separate piece of paper by tracing.

    a. Describe your strategies (other than random trial and error) for finding all possible shapes. Try to explain your strategies so that someone reading this description via email could understand it.
    b. Name each shape. If you know of a mathematical name, use it. If not, make up a name that is related in some way to the shape.
    c. Describe your other observations from this exploration: patterns, discoveries, conjectures, and questions.

3. *The two small triangles and the parallelogram*   With the same restrictions, make as many shapes as you can using only the two small triangles and the parallelogram. Record your findings on a separate piece of paper by tracing.

    a. Describe your strategies (other than random trial and error) for finding all possible shapes. Try to explain your strategies so that someone reading this description via email could understand it.
    b. Name each shape. If you know of a mathematical name, use it. If not, make up a name that is related in some way to the shape.
    c. Describe your other observations from this exploration: patterns, discoveries, conjectures, and questions.

4. *Making connections*

    a. Your instructor will have facilitated the making of a table that shows the shapes made in Steps 1, 2, and 3. In some cases all three columns are filled, indicating that this shape (for example, the "same" rectangle) could be made in all three cases. In some cases there will be blanks, indicating that the group was not able to make a particular shape with the three tangram pieces they had. Your task here is to examine the blanks that you or your instructor select and either make the shape or prove that it cannot be made.
    b. The number of shapes that were made from the two small triangles and the parallelogram was greater than the number of shapes that could be made from the two small triangles and the medium triangle or from the two small triangles and the square. Explain why we get more shapes with the two small triangles and the parallelogram.

PART 4: Congruence with tangrams

1. We can make a trapezoid with the tangram's parallelogram and the two little triangles, and we can make a trapezoid with the tangram's medium triangle and the two little triangles. Are the two trapezoids congruent? How could you prove that they are or are not?

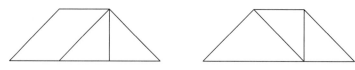

2. Are the members of each of these pairs of tangram figures congruent? Explain your reasoning.

a.

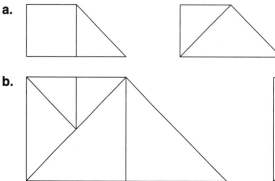

b.

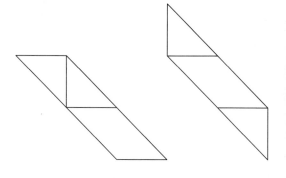

3. Three different words are commonly used by students when describing figures that have a certain amount of sameness: *same, congruent,* and *identical.* Some people argue that having three different words is picky. Recall that in Chapter 3 we found that many teachers prefer not to use *borrowing* and *carrying* but rather one word for both operations. What do you think? Do we need all three words? Why or why not?

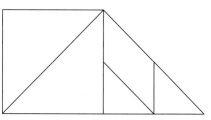

EXPLORATION 8.3   **Polyomino Explorations**

Polyominoes are geometric figures that are composed entirely of squares. You are probably familiar with two subsets of polyominoes: *dominoes* (made from two squares) and *tetrominoes* (made with four squares and made most popular by the Game Boy game Tetris®). The word *polyomino* was coined by Solomon Golumb, an American mathematician, in 1953, though puzzles with different kinds of polyominoes seem to have been around for centuries. You will find explorations with polyominoes in many elementary school textbooks, because they are such a versatile tool for exploring different mathematical ideas and developing spatial sense and problem-solving skills.

PART 1: Defining pentominoes

We will begin our explorations with *pentominoes* and communication. Below is a picture of two figures that are pentominoes and two figures that are not pentominoes.

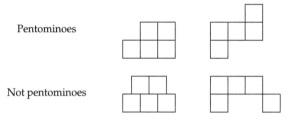

Pentominoes

Not pentominoes

1.   Write down your definition of *pentomino*.
2.   Exchange your definition with another student or read your definition to your group. Considering each definition as a first draft, identify any confusing or ambiguous words or phrases. Summarize the discussion of your definition. That is, identify any confusing or ambiguous words or phrases and explain why they were perceived to be confusing or ambiguous.
3.   Write your second draft of your definition of *pentomino*.

PART 2: How many polyominoes?

Note: In this exploration, you will create two sets of polyominoes—tetrominoes and pentominoes—using two or three pages of the Polyomino Grid Paper provided at the end of the book. Save your tetromino and pentomino sets for use in other explorations in Chapters 8 and 9.

   All dominoes have the same shape. If you look at the two dominoes below at the left, you can see that the second domino is the same shape as the first; it has simply been rotated 90 degrees. When we get to *trominoes*, made with three squares, the matter is still rather simple. There are two different trominoes, as shown in the figure below at the right.

Dominoes   Trominoes

1.   When we get to tetrominoes, made with four squares and popularized by the game Tetris®, there are a few more.
   a.   Determine how many different tetrominoes there are. Show your solutions by cutting out your different tetrominoes from the Polyomino Grid Paper at the end of the book.

    b. Share your solutions with members of your group. Write down any differences or insights—something you learned from the discussion.

2. When we search for all tetrominoes, a more or less random trial-and-error strategy is often sufficient. However, when we move to pentominoes, the number of possibilities is great enough so that it pays to be more systematic and less random.

    a. How many "different" pentominoes can be made? Show your solutions by cutting out your different pentominoes from the Polyomino Grid Paper at the end of the book. Describe any tools other than random trial and error that you used in your search.

    b. Once you think you have found all of the pentominoes, explain why you think you have them all.

    c. As a class, come up with a name for each different pentomino. This is necessary to make communication easier.

3. When we examine hexominoes, the number of possibilities increases even further. How many different hexominoes can you find? (You can use a copy of "Other Base Graph Paper" at the back of the book to record your examples of hexominoes). Because this is a mathematics course and we want your problem-solving tools to develop, the expectation is that you will use problem-solving strategies other than random trial and error. Think about this problem first. Next describe your plan of attack. Then do the problem, thinking and reflecting as you work on it. You might want to use the divided-page problem-solving format. At the end, summarize your strategies.

PART 3: Classifying pentominoes

1. Take your set of pentominoes and divide them into 2 or more subsets so that the members of each subset are alike in some way.

2. List the members of each subset.

3. Define each subset. That is, what are the membership criteria for the subset?

4. Repeat this process as many times as your instructor asks.

PART 4: A pentomino game

Materials:
- Your set of pentominoes (made in Part 2)
- Use a copy of the Polyomino Grid Paper at the end of the book to make a chessboard (8 × 8 square). Each square of your pentominoes is the same size as each of the 64 squares on the board.

Rules for competitive version

1. Players take turns placing one pentomino on the board.

2. The winner is the last person who can place a pentomino on the chessboard so that it does not extend beyond the board and does not lie on top of a previously played pentomino.

1. Play the game several times. Describe any strategies you discovered.

2. Play a cooperative version: Find the set of moves that results in the smallest number of pentominoes being placed on the board.

    a. Show your solution.

    b. Describe any strategies you discovered.

PART 5: Challenges

Use the tetromino and pentomino sets that you made in Part 2.

For Steps 1–3, take out the grids on pages 229–231. In each step, you will use a group of tetrominoes or pentominoes to fill the white spaces on the given grid. Record your solutions on the girds.

Note: In each of the challenges, describe strategies you used beyond guess–check–revise.

1. Using all five tetrominoes, completely fill grids (a), (b), and (c) on pages 229–230.
2. Use six copies of the L-shaped tetromino piece to fill the grid at the bottom of page 230.
3. Using four different petominoes, fill grids a and b on page 231.
4. Selecting among your set of pentominoes, solve the following puzzles.

   a. Use 4 different pentominoes to make a 4 × 5 rectangle.
   b. Use 6 different pentominoes to make a 3 × 10 rectangle.
   c. Use 5 different pentominoes to make a 5 × 5 square.

5. Using your whole set of 12 pentominoes, arrange the pentominoes so that the space enclosed by the pentominoes is as large as possible.
6. This challenge comes from a story, probably fiction, that goes like this. The son of William the Conqueror and the dauphin of France were playing a game of chess. At one point, the dauphin became angry and threw the chessboard at William's son. The board broke into thirteen pieces, 12 different pentominoes and 1 tetromino. Make a chessboard with the 12 different pentominoes and 1 tetromino.
7. Make up your own challenge. Describe your thinking process and a solution.

Grids for EXPLORATION 8.3, PART 5: Challenges

1. a.

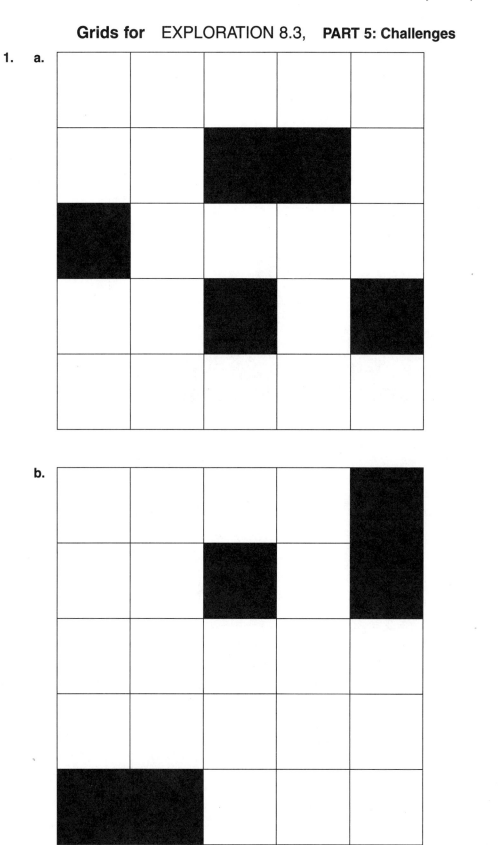

    b.

c.

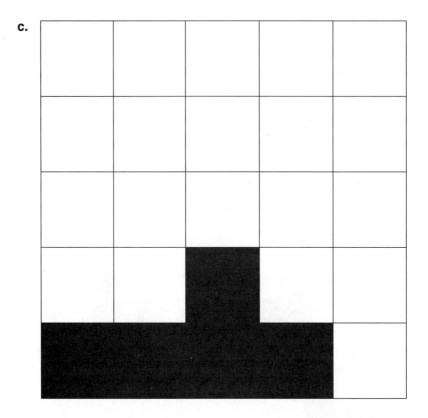

2.

Grids for EXPLORATION 8.3, **PART 5: Challenges**

3. **a.**

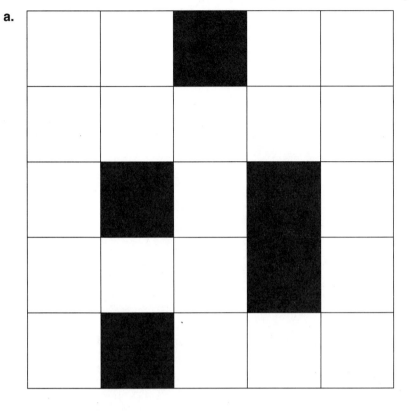

b.

PART 6: Folding polyominoes into cubes

1. Which of the pentominoes can be folded to make a box with no lid (that is, an open-ended cube)?
   a. Predict which ones will make a box (with no lid) when folded.
   b. Meet with your partner(s). Compare your predictions and reasoning.
   c. Determine which pentominoes do make boxes. If there were any that did make boxes that you didn't predict or any that didn't make boxes that you thought would, reflect on your reasoning. That is, now that you know the answers, can you add to your spatial reasoning tools—describe new tools or refine old ones?
   d. Identify any characteristics that distinguish those pentominoes that do make a box from those that do not make a box. Can you develop a function for describing which pentominoes will make boxes?

2. Which hexominoes can be folded to make a box with a lid (that is, a cube)? As you proceed on this problem, you should get to a point where you can rule out some possibilities just by looking. Describe, to the best of your ability, your "geometry sense" that enables you to sense "Yes, this one looks possible" or "No, not this one." For example, look at the two hexominoes below. My sense is that most of you would not predict that the one at the left would fold into a box and that most of you would predict that the one at the right would.

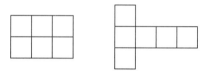

Nets

A *net* is a diagram showing what a hollow three-dimensional object would look like if it were cut along several edges, unfolded, and laid flat.

3. In each problem below, you are given one picture of a cube with three faces showing. From this information, determine which of the nets at the right is possible. In some cases, more than one of the nets are possible. Briefly describe your reasoning.
   a.
   b.

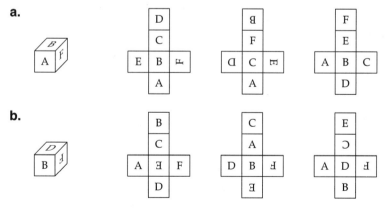

c.

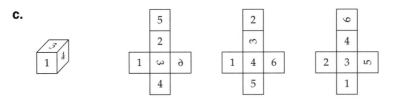

4. In each problem below, you are given three pictures of a cube. From this information, determine which letters are on opposite faces of the cube.

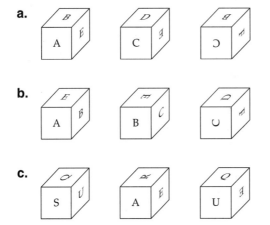

SECTION **8.1** **EXPLORING BASIC CONCEPTS OF GEOMETRY**

In order to work effectively with geometric ideas, we need to have a foundation of postulates, definitions, and theorems. As you may recall from high school geometry, the Greek contribution to geometry and mathematics is enormous. Though this textbook does not attempt to replicate the kind of formal geometric work you did in high school, I do believe it is helpful for elementary teachers to explore concepts and ideas that students will further examine in high school in order to have a better sense of the connection between the explorations students do in elementary school and the geometric knowledge they need to bring to middle school and high school.

EXPLORATION 8.4 **Seeing Patterns and Making Predictions**

This exploration contains several activities intended to increase your ability to think systematically, to see patterns, and to develop and test hypotheses that come from your observations, and to help you to see that algebra and geometry are not "disconnected" mathematical fields.

PART 1: How many regions?

When 2 lines in a plane intersect, 4 distinct regions are formed.

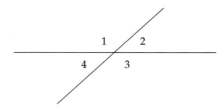

1. When 3 lines intersect in a plane, what is the maximum number of regions that can be formed?
2. When 4 lines intersect in a plane, what is the maximum number of regions that can be formed?
3. When 5 lines intersect in a plane, what is the maximum number of regions that can be formed?
4. When n lines intersect in a plane, what is the maximum number of regions that can be formed?

PART 2: How many lines?

If we take any 3 points that are not on the same line (noncollinear points), there are exactly 3 lines that can be constructed that will go through 2 of the 3 points. Thus, we can say that 3 noncollinear points determine 3 lines; that is, once you make the 3 points, those 3 lines are determined.

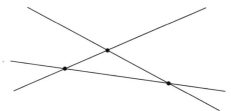

In a similar fashion, we find that any 4 noncollinear points will determine 6 lines. That is, regardless of where the 4 points are, as long as they are noncollinear, 6 lines are determined. See the following figures.

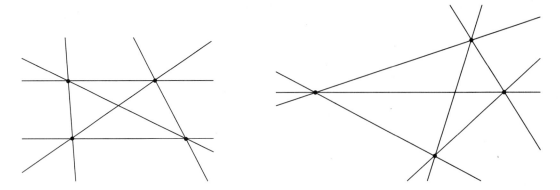

1. How many lines are determined by 5 noncollinear points?
2. How many lines are determined by 6 noncollinear points?
3. Continue this exploration with 7 noncollinear points, 8 noncollinear points, and so on, until you see patterns that enable you to answer the following question: How many lines are determined by *n* noncollinear points? Your response to this last question needs to include the answer, a summary of how you determined the answer, and your justification of your answer (that is, explain to the reader why you believe your answer is correct).

SECTION   **8.2**   **EXPLORING TWO-DIMENSIONAL FIGURES**

As you may already know, humans have been fascinated by various two-dimensional patterns for thousands of years. Many walls, ceilings, and floors have patterns. Patterns pervade many art forms. Understanding the properties of shapes and relationships between shapes enables the artist, the architect, and others to make interesting patterns and to design useful and appealing objects.

EXPLORATION 8.5   **What Makes a Shape a Shape?**

You will find variations of this exploration in many K–12 textbooks for several reasons. First, it develops good communication and classification skills. Second, it helps students to appreciate the need to have clear definitions in mathematics. Mathematicians stress the importance of using precise language when we are doing and discussing mathematics. Many students do not fully understand this need for precision and clarity and thus interpret their teachers' emphasis on precision as being "picky" or "technical" when, in fact, this push for precision comes from the desire to reduce confusion and ambiguity. Finally, most students find these explorations fun, like puzzles.

PART 1: New shapes

1.   In the group at the left below, you find four Wankas. None of the figures in the middle group are Wankas. Your task is to determine which of the four figures in the group at the right are Wankas.

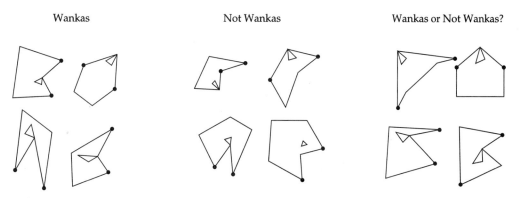

Wankas   Not Wankas   Wankas or Not Wankas?

    a.   Think about how you plan to go about determining which of the figures are Wankas. Then discuss your strategy with other members of your group. Describe your strategy.

    b.   Now, determine which of the figures in the group at the right are Wankas. Justify your choices.

    c.   Discuss your results with other members of your group and then with the whole class. If you got any wrong, describe changes and/or refinements in how you go about the process that you can use in the next step.

    d.   Write a definition for Wankas.

    e.   Give this definition to someone who is not taking this class and ask that person to draw two or more Wankas from your definition. Describe the results.

2. In the group at the left below, you find four Kiwis. None of the figures in the middle group are Kiwis. Determine which of the figures at the right are Kiwis.[1]

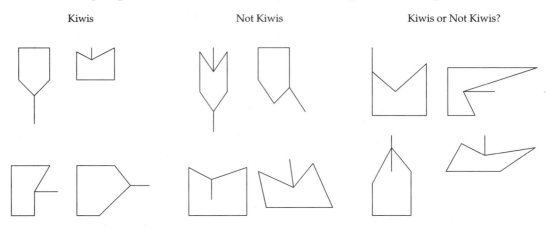

Kiwis   Not Kiwis   Kiwis or Not Kiwis?

a. Determine which of the figures in the group at the right are Kiwis. Justify your choices.
b. Discuss your results with other members of your group and then with the whole class. If you got any wrong, describe changes and/or refinements in how you go about the process that you can use in the next step.
c. Write a definition for Kiwis.
d. Give this definition to someone who is not taking this class and ask that person to draw two or more Kiwis from your definition. Describe the results.

3. In the group at the left below, there are four figures that are Xs. None of the figures in the middle group are Xs.

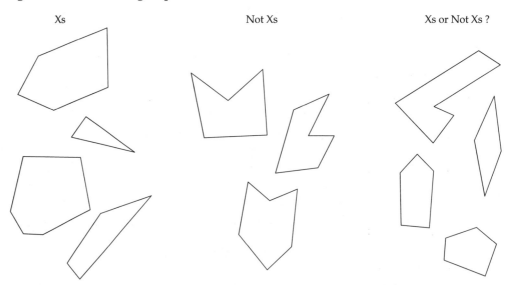

Xs   Not Xs   Xs or Not Xs ?

a. Determine which of the figures in the group at the right are Xs. Justify your choices.
b. Discuss your results with other members of your group and then with the whole class. If you got any wrong, describe changes and/or refinements in how you go about the process that you can use in the next step.
c. Write a definition of Xs.

[1] This question was adapted from one developed by a student of mine, Andrea Logue.

d. Give this definition to someone who is not taking this class and ask that person to draw two or more Xs from your definition. Describe the results.

4. In the first group at the left below, there are three figures that are Ys. None of the figures in the middle group are Ys.

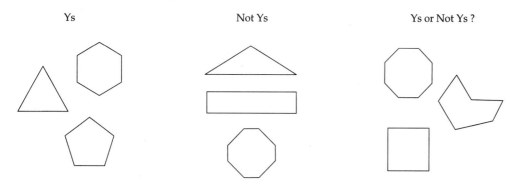

Ys   Not Ys   Ys or Not Ys ?

a. Determine which of the figures in the group at the right are Ys. Justify your choices.
b. Discuss your results with other members of your group and then with the whole class. If you got any wrong, describe changes and/or refinements in how you go about the process that you can use in the next step.
c. Write a definition of Ys.
d. Give this definition to someone who is not taking this class and ask that person to draw two or more Ys from your definition. Describe the results.

PART 2: 3RITs

In Chapter 9, we will explore the mathematics of quilting. One of the basic building blocks of many quilt patterns is the humble right isosceles triangle. The figures that follow include a "3RIT"—that is, a figure made from 3 right isosceles triangles.

3RIT   Not 3RIT

1. *Defining 3RITs*
    a. Discuss with your partner(s) your present thinking about what a 3RIT is. Draw more 3RITs and/or not 3RITs until you all agree on what a 3RIT is.
    b. On your own, write a precise definition of a 3RIT.
    c. Compare definitions with a partner with respect to accuracy and clarity.
    d. Modify your definition, if needed, so that it is accurate and clear.
2. *Making 3RITs*
    a. Make as many different 3RITs as you can. Sketch (or tape or glue) them on a separate piece of paper.
    b. Compare results with members of your group.

3. Are the two 3RITs below different or not? Discuss this question until your group arrives at a consensus. Explain why you believe they are or are not different.

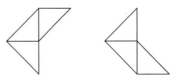

PART 3: 4RITs

1. *Making 4RITs*
    a. Make as many different 4RITs as you can. Sketch (or tape or glue) them on a separate piece of paper.
    b. Compare strategies and results with members of your group. Note any strategies mentioned in the discussion that you didn't use but that you liked. Also note any 4RITs that you didn't make.

2. *Classification*
    a. Divide your set of 4RITs into two or more subsets so that the members of each subset have at least one characteristic in common. Name each subset (feel free to be creative), and describe the defining characteristic(s) of that subset. Sketch (or tape or glue) the members of each subset on a separate piece of paper.
    b. Compare your results with members of another group.

EXPLORATION 8.6   **Exploring Triangles and Quadrilaterals with Grids**

As you have seen many times before in the text and in the explorations, different representations of a mathematical idea or problem can have interesting results. In Chapter 5, you saw how different manipulatives illustrated different aspects of the concept of fractions. In this exploration, we examine two (of many) possible grids that can be used to make geometric shapes. Quilters use different kinds of grids. Designs for expensive crystal are made by superimposing different grids on one another.

Consider the following types of triangles and quadrilaterals. Try to make each figure on the geoboard dot paper and isometric dot paper. If you can make the figure, do so and explain why your figure is an example of the specified triangle or quadrilateral. If you cannot make it, explain why you think it is impossible to make the figure on that type of dot paper.

1. Acute scalene triangle
2. Right isosceles triangle
3. Obtuse isosceles triangle
4. Equilateral triangle
5. Trapezoid
6. Kite
7. Parallelogram
8. Rectangle
9. Rhombus
10. Square
11. Square with no sides parallel to the sides of the paper

SECTION   **8.3**   **EXPLORING THREE-DIMENSIONAL FIGURES**

How do you represent a three-dimensional object on paper? Why is this question important? Think about these two questions before reading on. . . . 🗒

When we make complex objects, whether they be cars or houses or sculptures, they have to be designed first. Although scale models are often made, the actual construction is done from blueprints or drawings—that is, two-dimensional representations of the object. This ability to translate both ways between three-dimensional objects and two-dimensional representations of them is important in many parts of our life, yet this part of mathematics has been underemphasized in school in favor of more computational aspects of mathematics. In your work with children, it will be important that you provide them with opportunities to develop what we call spatial visualization abilities. More and more textbooks and other curriculum materials are emphasizing spatial visualization. It is very likely that you will do explorations like the ones below, at a simpler level, with your future students. The following explorations provide you an opportunity to work on these concepts and ideas and to do so in a way that should be both enjoyable and challenging.

EXPLORATION 8.7   **3D to 2D Explorations**

Here you will explore different possible ways to represent a three-dimensional object on paper. One of the beauties of this exploration is that there are many different ways to solve the problems, not just "the right way."

PART 1: How many cubes?

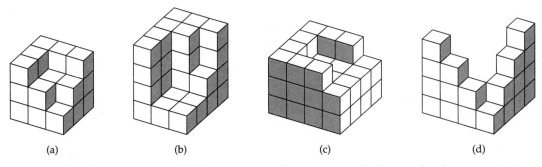

(a)   (b)   (c)   (d)

1.   **a.** Look at structure (a). How many cubes were needed to make the structure? Work on this alone and briefly note your work.
     b. Compare strategies with other members of your group. Note any strategies that you like.

2.   **a.** Look at structure (b). How many cubes were needed to make the structure? Work on this alone and briefly note your work.
     b. Compare strategies with other members of your group. Note any strategies that you like.

3.   **a.** Look at structure (c). How many cubes were needed to make the structure? Work on this alone and briefly note your work.
     b. Compare strategies with other members of your group. Note any strategies that you like.

4.   Write the minimum and maximum number of cubes that could be contained in structure (d) if it is viewed from "ground level." Justify your answer.

PART 2: Giving directions

A very major role that mathematics has played in the development of much technology is a process that is called mathematizing—that is, taking a situation or problem and using mathematics to understand that situation or problem better. Here you will grapple with how to give directions so that someone else can make a copy of a three-dimensional figure that you have constructed. If we make a block building from cubes (as in Part 1) and someone else wants to make that same building, that person can come to us and make a copy of that building right next to ours. If that is inconvenient, we could take a photograph, and the person could construct the building from the photograph. However, as you may have realized in Part 1, photographs can't show everything. Of course, we could photograph the building from all sides. That would solve the problem if we have a simple building (like the ones in Part 1), but it would not be useful for more complex buildings.

1.   **a.**   Make a building out of blocks and develop a way to communicate how to build that building to another group. Your communication may consist of words and/or diagrams.

      b.   Give your directions to another group and have them construct the building from your directions. Note any parts that they have difficulty with—that is, that they find confusing or ambiguous.

      c.   Switch roles, so that your group constructs a building from the other group's directions. Note any parts that you have difficulty with—that is, that you find confusing or ambiguous.

      d.   Repeat this process until you have developed a method that will enable another group to build successfully an exact copy of your block building.

2.   Listen to methods developed by other groups. Note any ideas or comments that you heard that you liked.

3.   Imagine that you are an elementary teacher and that students all over the country are making such buildings and comparing their work with that of other students. If you could express your choice among the different methods (via fax), which would you choose? Your response needs two parts: (1) an analysis of the pros and cons of each method, and (2) your choice and justification of your choice.

4.   Write your directions for making the buildings shown below.

    a.   **b.** 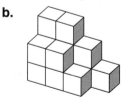   **c.**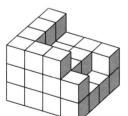

PART 3: Making drawings with isometric dot paper

We have explored different ways to communicate how to make block buildings. Now we will explore how to draw them. A special kind of dot paper called *isometric dot paper* (found at the end of the book) has been created to make such drawings easier. Make several copies of that page.

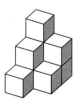

1. Draw the figure shown above on your dot paper. Many students find that it takes several rounds of guess–check–revise before their picture matches the one above. Once you succeed:

   a. Describe strategies that you discovered while copying the shape.

   b. Note other observations, insights, and questions (including what-if questions).

2. Practice drawing on dot paper by drawing the figures below and/or making block buildings and then drawing them. After these further explorations:

   a. Describe new strategies that you discovered while making your drawings.

   b. Note other observations, insights, and questions (including what-if questions).

3. **a.** Draw a building that contains between 10 and 15 cubes and is a "challenge" to draw—for example, a building with fewer cubes in a back row than in a front row.

   b. Write the directions for making that building.

   c. Explain any difficulties and how you overcame them.

PART 4: Interlocking cubes

1. What if you could make buildings with interlocking cubes, like the one below, for example?

   a. Make a complex building with interlocking cubes, and write directions for making the building, using any of the methods already devised or a new method (which could be a modification of previous methods). Have another group make the building using your directions.

   b. Report the results. If it wasn't successful, explain the glitches and how you solved the glitches. Then repeat Step 1(a).

2. What if the buildings were not restricted to cubes? Make a set of directions for one of the buildings below.

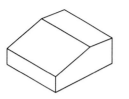   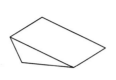   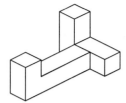

EXPLORATION 8.8   **Understanding Solids**

You found that each two-dimensional figure that has a name also has certain clearly stated attributes and that there are various relationships among two-dimensional figures—for example, squares are closely related to both rectangles and rhombi. These observations are also true for three-dimensional objects.

PART 1: Describing solids

This first part is similar to the introductory geoboard exploration: How can you describe to someone else, using only words, what you are holding in your hand? Students will work in pairs.

1.   Your instructor will show an object to the students who are facing the front. Using only words, these students will try to describe the object so that their partner can determine what it is. Afterwards, discuss the descriptions. If the students facing the back were able to decode the directions, great. If there were problems, analyze them, and then discuss how to revise the description to address the problems raised.
2.   Switch roles and repeat the process with another object.
3.   After the whole-class discussion, record your observations, your insights, and any questions that you still have.

PART 2: Classifying solids

This second part focuses on understanding different relationships among three-dimensional figures.
    Examine the objects (or pictures of objects) you have in front of you. As we have done in several other explorations, place the objects into two or more subsets so that the elements in each subset are alike in some way.

1.   For each subset, describe the common attributes or properties, give a name to the subset (if you don't know of a mathematical name, make up a name), and list the members of the subset.
2.   Repeat this process several times.
3.   What did you learn from this exploration?

PART 3: Nets

You will be asked to make nets for different figures using posterboard, ruler, and scissors (or exacto knives). A net is a diagram showing what a hollow three-dimensional object would look like if it were cut along several edges, unfolded, and laid flat. For example, one net for the cube at the left is shown at the right.

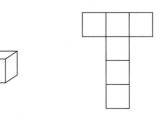

    Two notes before you begin:
- Before you cut anything, you have to have the complete net drawn on the paper.
- The net cannot contain separate pieces. For example, the net above represents a

valid way to make a cube. Cutting six squares of the same size and taping them together would be an invalid way.

For each figure shown in Steps 1–4, do the following:

a.   Make a net for the figure.
b.   Describe strategies that went beyond simple trial and error.
c.   Do you think there are other possible nets, or just this one? If you think there are others, sketch them.

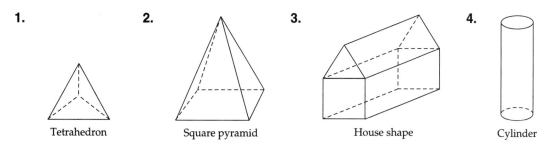

| 1. | 2. | 3. | 4. |
| Tetrahedron | Square pyramid | House shape | Cylinder |

Extensions

1.   Repeat Part 1 of this exploration with new objects.
2.   Describe changes in your way of describing objects that you learned in this exploration.

EXPLORATION 8.9   **Cross Sections**

One important spatial visualization skill is being able to imagine what happens when physical objects are changed or moved. Examples include being able to imagine how a new arrangement of furniture will work in a living room and being able to figure out whether all of a set of objects will fit into a given space (such as a suitcase or the trunk of a car).

Here, we will explore the notion of cross sections. Consider the cube at the right. If we sliced it in half (as though a knife made a vertical slice), what would the newly exposed face (that is, the cross section) look like? If we sliced it at an angle, what would the new face look like now?

PART 1: Predictions

For each figure in the table on pages 249–251, first predict what the sliced face will look like. Then check your prediction either by making a clay figure and slicing it or by some other means.

PART 2: Descriptions of intersections

In Part 1, you were given a picture representing a specific cross section of a solid object. What if you had no picture?

1.   Describe an object and an intersecting plane, using only words.
2.   Have another group describe what slice they would make, on the basis of your description. If they are able to decode your directions accurately, great. If there are problems, analyze them, and then revise the description to address the problems raised.

Table for EXPLORATION 8.9, **Cross Sections**

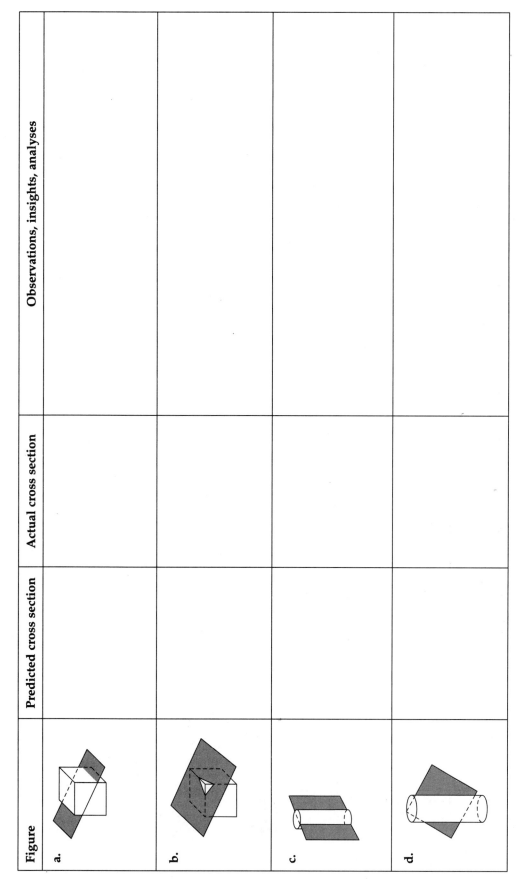

Figure	Predicted cross section	Actual cross section	Observations, insights, analyses
a.			
b.			
c.			
d.			

Table for EXPLORATION 8.9, **Cross Sections**

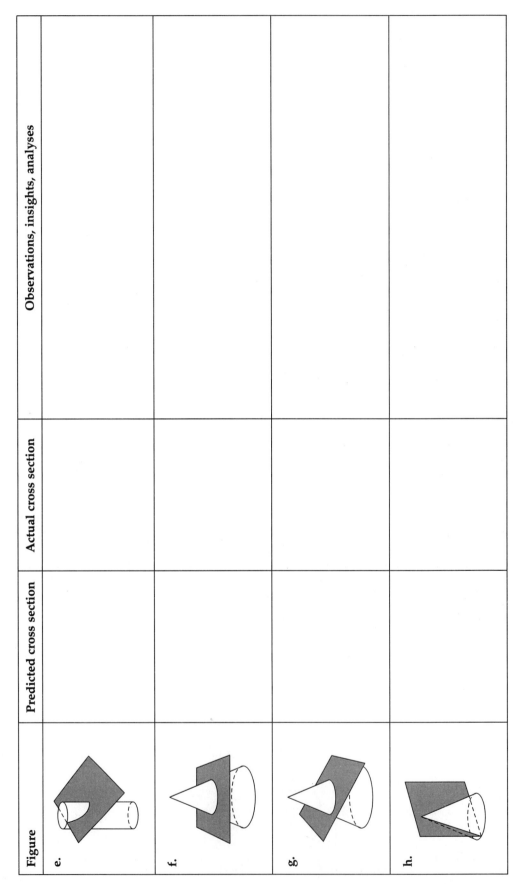

Figure	Predicted cross section	Actual cross section	Observations, insights, analyses
e.			
f.			
g.			
h.			

Table for EXPLORATION 8.9,   **Cross Sections**

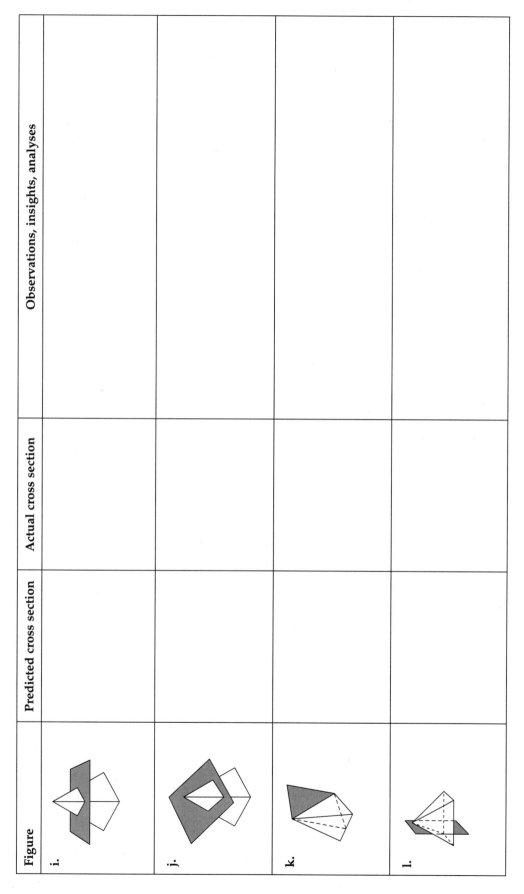

EXPLORATION 8.10   **Relationships Among Polyhedra**

Leonard Euler discovered a relationship among the number of vertices, the number of edges, and the number of faces of any *polyhedron*. Rather than present this relationship, I ask you to look at a variety of polyhedra following.

PART 1: Collecting data and making hypotheses

1.   Count and record in the table on page 255–256 the numbers of vertices, edges, and faces in each polyhedron.
2.   What patterns do you see, and what observations or hypotheses do you have? Record them in the table. If you keep your mind open and use the problem-solving tools you have been developing, you will discover more than Euler's formula.

PART 2: Discoveries

1.   Describe the relationship you found among the numbers of vertices, faces, and edges of any polyhedron.
2.   Describe any significant actions that you took that facilitated this discovery.

Table for EXPLORATION 8.10, Relationships Among Polyhedra

Polyhedron	Number of vertices	Number of faces	Number of edges	Patterns, observations, hypotheses
a.				
b.				
c.				
d.				

Table for EXPLORATION 8.10, **Relationships Among Polyhedra**

Polyhedron	Number of vertices	Number of faces	Number of edges	Patterns, observations, hypotheses
e.				
f.				
g.				

9

Geometry as Transforming Shapes

In Chapter 8, we explored geometry from the perspective of shapes and properties. We found that the theme of composition and decomposition, which we first discovered with numbers, also extends to geometry. That is, by decomposing shapes, we can better understand their properties and their connections to other shapes. In this chapter, we are going to focus on *transformations* — that is, on what happens when we move shapes or modify shapes.

As in Chapter 8, the first three explorations use geoboards, tangrams, and polyominoes to explore topics from throughout the chapter.

EXPLORATION 9.1   Geoboard Explorations

PART 1: Slides, flips, and turns

1. **a.** Write directions for moving each of the figures from position A to position B.
    b. Exchange directions with someone else.
    c. On the basis of the feedback you receive, either keep your response to part (a) or write a second draft.

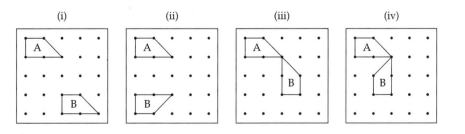

(i)   (ii)   (iii)   (iv)

2. **a.** Write directions for moving each of the figures from position A to position B.
   b. Exchange directions with someone else.
   c. On the basis of the feedback you receive, either keep your response to part (a) or write a second draft.

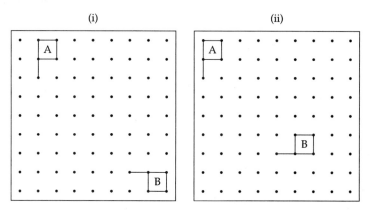

3. Use the figures on page 261. In each problem, reflect the polygon across the given line and sketch the reflection of the figure.

4. For each of the following, use a geoboard or geoboard dot paper.

   a. Make a figure that will look the same when we turn the geoboard 90 degrees (a 1/4 turn), 180 degrees (a 1/2 turn), or 270 degrees (a 3/4 turn).
   b. Make a figure that will look the same when we turn the geoboard 180 degrees but different when we turn it 90 degrees.
   c. Make a figure that will look the same when we flip the geoboard across a horizontal line (so that the figure is turned over). *Note:* In this case, you have to imagine a transparent geoboard.
   d. Make a figure that will look the same when we flip the geoboard horizontally but different when we flip it vertically.
   e. Make a figure that will look the same no matter how we flip or turn the geoboard.

PART 2: Symmetry

For Steps 1–3, use the figures on page 263.

1. Determine whether each figure is symmetric. If so, describe the line(s) and/or rotation(s).
2. Finish figures a and b so that they have at least one line of symmetry. Finish figures c and d so that they have rotational symmetry.
3. Make your own figures for a classmate to test for symmetry.
4. Make a square on geoboard dot paper.

   a. Modify one side. Now modify the opposite side so that the figure still has symmetry.
   b. Modify another side. Now modify the opposite side so that the figure still has symmetry.
   c. Repeat parts (a) and (b) with other squares and different modifications. What generalizations can you make?
   d. Repeat parts (a) and (b) with other symmetric polygons that have an even number of sides. What generalizations can you make?

5. Make a square on geoboard dot paper.

   a. Modify one side. Now modify the side to the right so that the figure still has symmetry. Repeat the process, making the same modification each time, until all four sides of the square have been modified.

b. Repeat part (a) with other squares and different modifications. What generalizations can you make?

c. Repeat this with other symmetric polygons with an even number of sides. What generalizations can you make?

PART 3: Similarity

1.   Use the figures on page 265 and several sheets of geoboard dot paper. For each figure, make a larger copy of the shape on a sheet of geoboard dot paper so that the original and the larger shape are "similar." Briefly describe how you know that they are similar.

2.   Compare your results with those of other members of your group.

3.   Draw your own shapes on geoboard dot paper, and then make larger copies as in Step 1.

4.   Write a definition of *similar*. That is, develop a definition that someone else could use to determine whether two figures are similar. Another way of this thinking about this question is to determine what relationship(s) a smaller and larger figure must have in order to be similar.

Figures for EXPLORATION 9.1, **PART 1: Slides, flips, and turns**

3. **a.**   **b.**

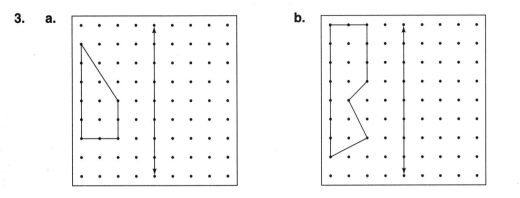

c.   **d.**

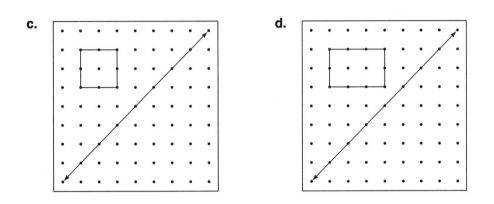

Figures for   EXPLORATION 9.1,   PART 2: Symmetry

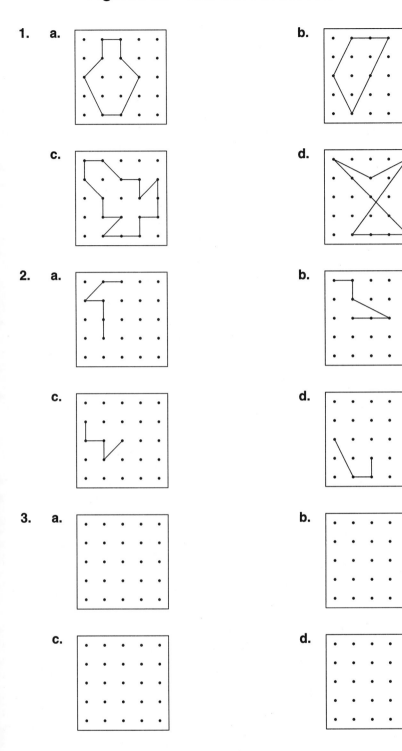

1.   a.   **b.**

c.   **d.**

2.   a.   **b.**

c.   **d.**

3.   a.   **b.**

c.   **d.**

Figures for EXPLORATION 9.1, **PART 3: Similarity**

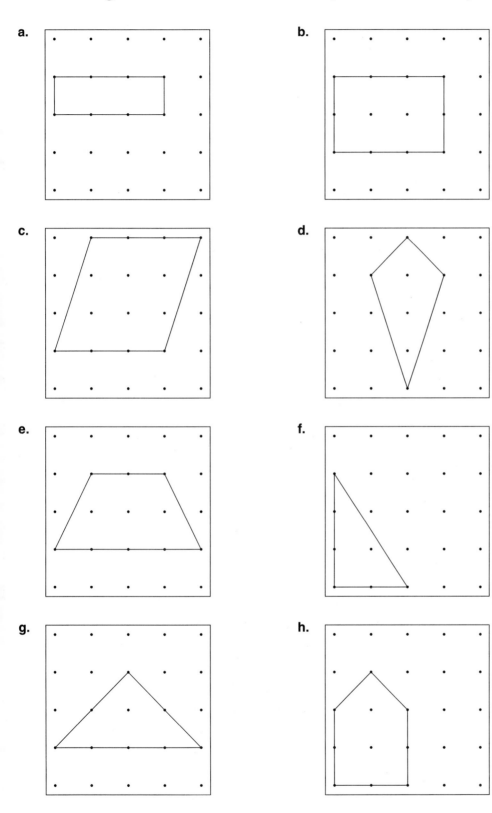

EXPLORATION 9.2   **Tangram Explorations**

PART 1: Symmetry

1. Make different shapes from your tangram pieces so that each overall shape has one line of symmetry.
2. Make different shapes from your tangram pieces so that each overall shape has two lines of symmetry.
3. Make different shapes from your tangram pieces so that each overall shape can be rotated and still look the same. Specify the nature of the rotation symmetry of the shape.

PART 2: Similarity

Rectangles

1. a. With your group, make as many different rectangles as you can, using the pieces from one tangram set. Make a copy of each rectangle by tracing it.
    b. Compare your results with those of another group (or the whole class). Add any new drawings of rectangles to your group's collection.

2. a. Separate the set of rectangles into subsets so that all the rectangles in a subset are similar to each other.
    b. Describe how you determined which rectangles are similar. Then write a definition or a rule for *similar* rectangles. This will be your first draft.
    c. Share your definition with other groups.
    d. If your definition/rule has changed, note the change and explain why you like this version better than your original version.

Trapezoids

3. a. Make as many different trapezoids as you can, using pieces from one tangram set. Make a copy of each trapezoid by tracing it.
    b. Compare your results with those of another group (or the whole class). Add any new drawings of trapezoids to your group's collection.

4. a. Separate the trapezoids into subsets so that all the trapezoids in a subset are similar to each other.
    b. Use your rule/definition from Step 2 to determine which trapezoids are similar. If it still works, move on. If it doesn't, explain why, and then modify your rule/definition.
    c. Share your definition with other groups.
    d. If your definition/rule has changed, note the change and explain why you like this version better than your original version.

Extensions

5. Can you extend your definition of similarity to any geometric figure? Make a tangram figure that is not a triangle or a quadrilateral. Now make a similar figure on another sheet of paper. Explain how you did it and justify your solution.
6. Irma says that all of the triangles that you can make with tangrams will be similar to one another. What do you think of Irma's conjecture? Explain your reasoning.

EXPLORATION 9.3   **Polyomino Explorations**

Use the tetromino and pentomino sets that you made in Exploration 8.3 (see page 225).

PART 1: Slides, flips, and turns with pentominoes

1.   There are four pairs of pentomino figures below. In each case, the top figure was transformed into the bottom figure by a flip or a turn. Describe what was done to the top figure to create the bottom figure. Use your pentomino pieces to check your answer.

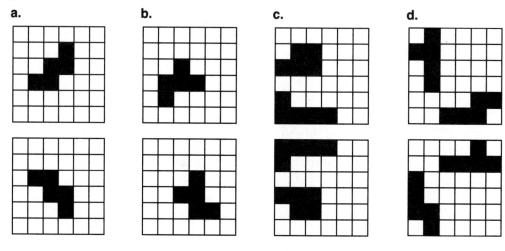

    a.   **b.**   **c.**   **d.**

2.   Describe what was done to the top figure to create the bottom figure.

    a.   **b.**   **c.**   **d.**

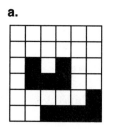
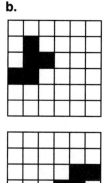
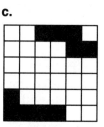
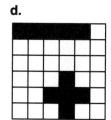
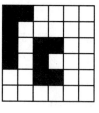

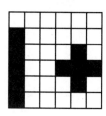

3.   Use the four grids on page 269. Make your own pentomino figures on the top grid. Then transform the figure by making any combination of flips or turns. Record the result on the bottom grid. Describe what you did. Give your problems to a friend and see whether your friend can guess your transformations.

Grids for EXPLORATION 9.3,
PART 1: Slides, flips, and turns with pentominoes

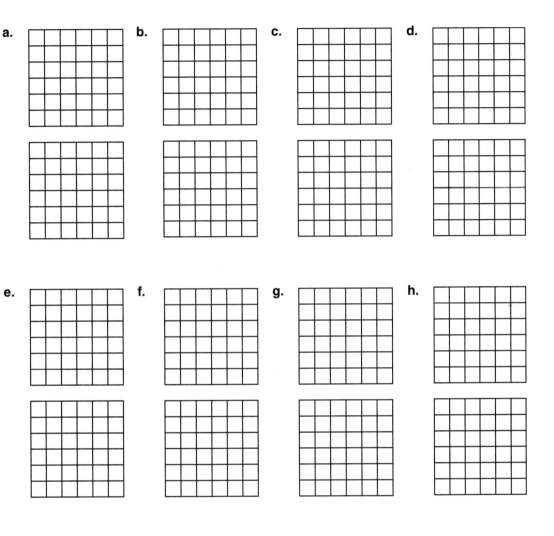

PART 2: Tessellations

1.   Which of the five tetromino pieces will tile the plane? Show your solutions. *Note*: Some of the pieces will tile the plane in more than one way; that is, they will make different patterns.
2.   Make a tessellation pattern with two or more of the tetromino pieces.

PART 3: Symmetry

Describe the symmetries of each of the 12 pentomino pieces.

PART 4: Similarity

Select the pentominoes that are shaped like the following capital letters: T, U, Z, W, and P.

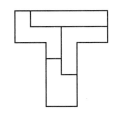

For each one of these five shapes, a similar figure can be made by using four other pentominoes (of any shape). The solution for T is shown.

1.   Take a few minutes to think about making a U, Z, W, or P. What ideas do you have beyond a random guess–check–revise?
2.   Share your thoughts with your partners. Together, make a figure similar to one of these pentominoes.
3.   At the most basic van Hiele level, the larger figures are similar in appearance to the smaller figures. However, if we analyze the smaller and larger figures, we can find mathematical relationships between the two figures. Take some time in your group to grapple with this concept of similarity and then draft a definition of similarity that you can then test on these figures.
4.   Share your definition with other groups.
5.   If your definition has changed, note the changes and explain why you like this version better than your original version.

SECTION   **9.1**   **EXPLORING TRANSLATIONS,**
REFLECTIONS, AND ROTATIONS

In this section, we will explore the effects of moving two-dimensional figures in various ways: by sliding the shape (called a *translation* of the shape), by flipping the shape (called a *reflection* of the shape), and by turning the shape (called a *rotation* of the shape).

EXPLORATION 9.4   **Reflections (Flips)**

In this exploration, we will focus on understanding what happens to a figure when it is reflected across a line and when it is reflected across more than one line. In doing so, you will not only learn more about reflections but also develop the ability to make and test predictions.

PART 1: Developing reflection sense

In the left part of the figure below, we see the effect of reflecting triangle *CAN* across a line. Triangle *C'A'N'* is called the reflection image of triangle *CAN*. In the right part of the figure, we see the effect of reflecting rectangle *PONY* across a diagonal line.

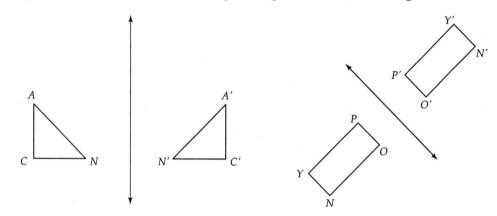

There are some similarities and some differences between the two reflections. In the first case, we reflected a triangle; in the second case, a rectangle. In the first case, the *line of reflection* was a vertical line; in the second case, the line of reflection was a diagonal line. In the first case, a single vertex was closest to the line of reflection; in the second case, one whole side of the figure was closest to the line of reflection. In this exploration, you will be asked to discover the common characteristics of all reflections. For example, we found that all rectangles share certain characteristics, such as four sides and four 90-degree angles.

1.   Take out page 277. For each of the figures at the top the page, do the following:
    (i)   Sketch lightly with a pencil where you think the reflection of each shape will be.
    (ii)   Then reflect the shape according to your instructor's directions or by the following method, which, in effect, has you make your own carbon copy: (1) Using either pencil or ink, trace one of the figures. (2) Fold the paper on the line of reflection. (3) Trace the figure again, this time on the opposite side of the paper.
    (iii)   Compare the actual reflection and your prediction. If your prediction was close, great! If it was not, take a moment to analyze your error. Why was

your prediction wrong? How wrong was it? What can you learn about the relationship between a figure and its reflected image? Write down your analysis!

2. After doing all six reflections in Step 1, describe your method for determining a reflected image. That is, what generalizations can you make about reflecting any figure across any line?

3. For each pair of figures at the bottom of page 277, do the following:

   (i) Sketch lightly with a pencil where you think the line of reflection is for each pair of shapes.

   (ii) Determine the actual line by folding the paper so that the congruent figures lie directly one on top of the other.

   (iii) Compare the actual line and your prediction. If your predicted line was close, great! If it was not, take a moment to analyze your error. Why was your prediction wrong? How wrong was it? What can you learn about the relationship between a figure and its reflected image? Write down your analysis!

4. After doing all six pairs of figures in Step 3, describe your method for determining a line of reflection. That is, what generalizations can you make about the line of reflection for a figure and its reflection?

PART 2: Two reflections

Now that you have examined the operation of reflecting a figure across a single line, let us examine the operation of reflecting a figure across two lines in turn.

1. We will first explore the effect of reflecting a figure across two lines that are parallel. Use the first figure on page 279.

   a. Predict the reflection image of trapezoid *STAR* if we first reflect it across line *l* and then reflect that image across line *m*. Sketch where you think the final image will be. Briefly explain your prediction.

   b. Now, do the actual reflections. If your prediction was way off, describe what you learned from this activity that will make your future predictions more accurate.

   c. Predict the reflection image of trapezoid *STAR* if we first reflect it across line *m* and then reflect that image across line *l*. Sketch where you think the final image will be. Briefly explain your prediction.

   d. Do the actual reflections. If your prediction was way off, describe what you learned from this activity that will make your future predictions more accurate.

2. a. On a sheet of blank paper, make a simple geometric figure and two parallel lines *l* and *m*. First, reflect this figure across the line *l* and then line *m*. Then reflect it across line *m* and then line *l*.

   b. Repeat this process with other figures until you feel confident about answering part (c).

   c. Describe your conclusions about the effects of reflecting a figure across two parallel lines.

3. You found that when we reflected trapezoid *STAR* across line *l* and then across line *m*, we did not get the same result as when we reflected it first across line *m* and then across line *l*. If we let X represent any figure, we can symbolize this observation as $Xlm \neq Xml$. From one perspective, reflection across a line is an operation on a figure. This is analogous to the operation of addition. In the case of addition, we are operating on numbers. When working with numbers, we found

that the operation of addition is commutative and that the operation of subtraction is not commutative.

a. What about the operation of reflection? Are there any cases in which this operation will be commutative? What do you think?

When mathematicians wish to investigate a phenomenon, they often work systematically, exploring different cases. Let us explore another case: What if the figure is between the two parallel lines? Where will the reflection image lie? In this case, will $Xlm = Xml$?

Use the second figure on page 279.

b. Predict the reflection image of trapezoid *STAR* if we first reflect it across line *l* and then reflect that image across line *m*. Sketch where you think the final image will be. Briefly explain your prediction.

c. Do the actual reflections. If your prediction was way off, describe what you learned from this activity that will make your future predictions more accurate.

d. Predict the reflection image of trapezoid *STAR* if we first reflect it across line *m* and then reflect that image across line *l*. Sketch where you think the final image will be. Briefly explain your prediction.

e. Do the actual reflections. If your prediction was way off, describe what you learned from this activity that will make your future predictions more accurate.

4. a. On a sheet of blank paper, draw two parallel lines *l* and *m*, and then draw a simple geometric figure between the two lines. First, reflect it across line *l* and then line *m*. Then reflect it across line *m* and then line *l*.

b. Repeat this process with other figures until you feel confident about answering part (c).

c. Describe your conclusions about the effects of reflecting a figure across two parallel lines when the figure is between two lines.

d. What do you think now? Do you think there is any scenario in which Xlm and Xml will be equal when X is reflected across two parallel lines? Explain your reasoning.

5. Let us now examine reflections across perpendicular lines. Use the third figure on page 279.

a. Predict the reflection image of trapezoid *STAR* if we first reflect it across line *l* and then reflect that image across line *m*. Sketch where you think the final image will be. Briefly explain your prediction.

b. Do the actual reflections. If your prediction was way off, describe what you learned from this activity that will make your future predictions more accurate.

c. Predict the reflection image of trapezoid *STAR* if we first reflect it across line *m* and then reflect that image across line *l*. Sketch where you think the final image will be. Briefly explain your prediction.

d. Do the actual reflections. If your prediction was way off, describe what you learned from this activity that will make your future predictions more accurate.

6. We found that reflection of trapezoid *STAR* across the two perpendicular lines was commutative. Do you think this will always be the case? For example, what if the base of the figure was not parallel to one of the lines of reflection?

   a. Describe your initial hypothesis and briefly explain your reasoning.
   b. Do a number of reflections on your own. Show this work. In each case, briefly explain why you picked the figure that you picked.
   c. Describe your present hypothesis and briefly explain your reasoning.

7. Is there any other transformation or combination of transformations that could achieve the same result as reflecting a figure across two parallel lines? State your thinking and support your statement.

8. Is there any other transformation or combination of transformations that could achieve the same result as reflecting a figure across two perpendicular lines? State your thinking and support your statement.

Figures for EXPLORATION 9.4, **PART 1: Developing reflection sense**

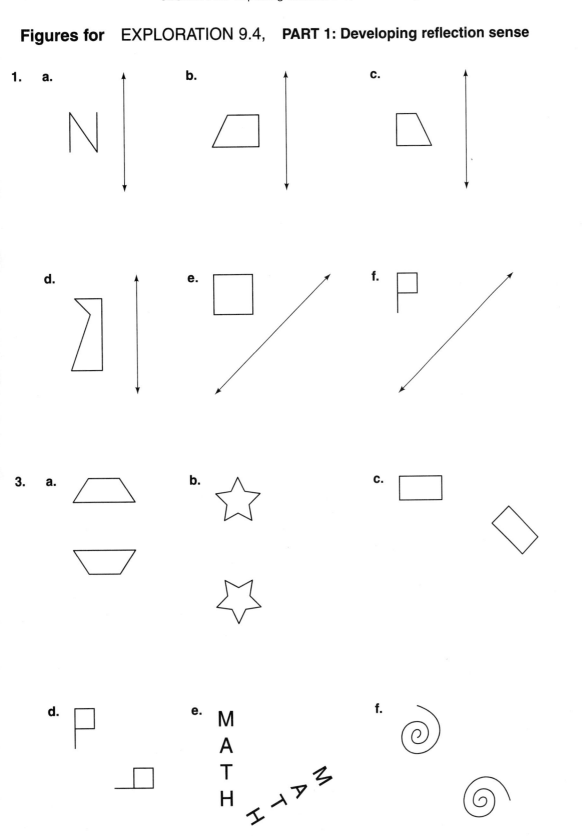

Figures for EXPLORATION 9.4, **PART 2: Two reflections**

1.

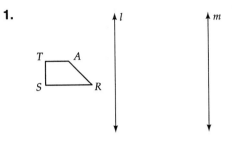

3.

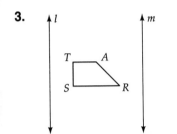

5.

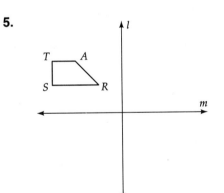

E X P L O R A T I O N 9.5   **Developing Rotation Sense**

1. For each of the figures on page 283, do the following:
    (i)   First, predict the image when the figure is rotated 180 degrees clockwise about the dot by lightly sketching your prediction in pencil on the sheet.
    (ii) Then do the rotation according to your instructor's directions or by the following method. Take a blank sheet of paper and trace each figure and dot (or simply make a copy of the page). Place one sheet above the other. With your pencil or pen firmly on the dot, turn the bottom sheet 180 degrees. Now trace the image.
    (iii) Compare the actual rotation and your prediction. If your prediction was very close, great! If it was not, take a moment to analyze your error. Why was your prediction wrong, how wrong was it, and what can you learn about the relationship between a figure and its rotated image? Write down your analysis!

2. After doing all 9 rotations, describe your method for determining a rotated image. That is, what generalizations can you make about rotating any figure 180 degrees clockwise? *Note*: you are welcome to make up new figures and do more exploring before answering this question.

3. Study the first six rotations (where the dot was not on the figure). What commonalities do you observe between the figure and its rotated image that are true in all six cases?

a.

b.

c. A

d.

e.

f. A

g.

h.

i. A

EXPLORATION 9.6   **Paper Folding**

In some of the problems below, you can begin with
a regular blank sheet of notebook paper. In other
cases, you need to begin with a square sheet of
paper. You can quickly make a square from a blank
sheet of notebook paper that measures $8\frac{1}{2}$ inches by
11 inches by folding one corner of the paper down
to the opposite side and then cutting off the excess,
as shown in the diagram at the right.

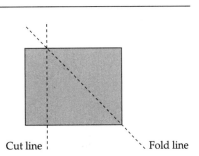

Cut line   Fold line

PART 1: Making geometric figures with paper folding

1. *Square*   Cut two strips of paper of equal
    width. Fold each strip onto itself. Insert one
    strip inside the other so that they interlock, as
    shown at the right. Cut off the excess paper,
    and you have a square (actually four squares
    if you unfold what remains and cut at the
    folds). Why are these figures squares?

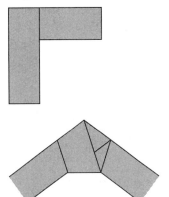

2. *Pentagon*   Cut a long, thin strip of paper.
    (This exercise will work with a thin strip cut
    from $8\frac{1}{2}$ by 11 inch notebook paper). Tie a
    regular overhand knot with the paper and
    carefully tighten the knot until your paper
    looks like the figure at the right. You may
    have to play with it for a bit. When you cut
    off the two excess pieces, you have a regular
    pentagon. What can you see from the folds
    that can help you to understand why the
    resulting figure is a regular pentagon?

3. *Hexagon* Cut two thin strips of paper. Again, this will work with strips cut from $8\frac{1}{2}$ by 11 inch notebook paper. Tie a square knot with the two pieces of paper. The figure at the left below shows the parts of each strip that are above and below the other strip. Be patient—it takes a couple readings to figure this out. Carefully tighten the knot; as you begin to tighten the knot, your strips should look like the figure at the right. (*Hint*: Make sure that the left and right end pieces are on top of each other as shown in the figure at the left.) What can you see from the folds that can help you to understand why the resulting figure is a regular hexagon?

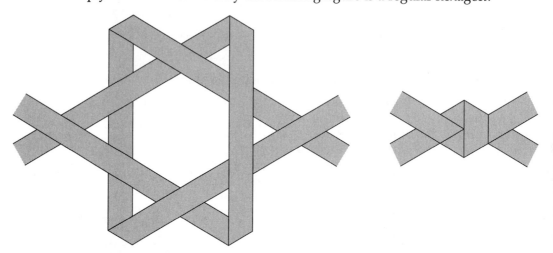

4. *A family of fives*
   a. Take a sheet of notebook paper and fold it in half as shown in Step 1.
   b. Find the midpoint of the bottom edge by folding. Bring the top left corner of the paper to the midpoint and fold, as shown in Step 2.
   c. Fold the bottom left corner across the fold line as shown in Step 3.
   d. Finally, fold the top edge across the fold line as shown in Step 4. Your result should look like the righthand figure in Step 4.

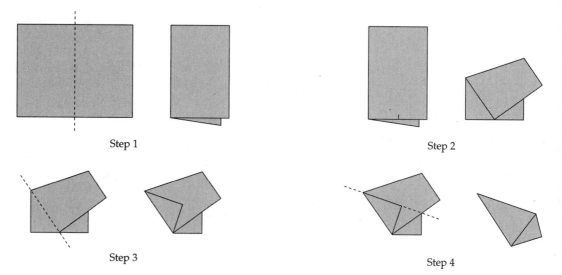

Step 1

Step 2

Step 3

Step 4

   e. Cut off the tip at an angle and then open up your paper.
   f. Repeat steps (a) through (e), making cuts of various sizes and angles at the tip.
   g. Present your results. Explain why each cut produces the kind of figure that it produces.

PART 2: Predicting what happens when you unfold the paper[1]

1. In the example shown below, a piece of paper is folded twice, and then an arrow is drawn in the upper right corner. This is a special type of paper: When the arrow is drawn, its image appears on each of the three layers below. When the paper is unfolded, the four arrows appear as shown.

Example: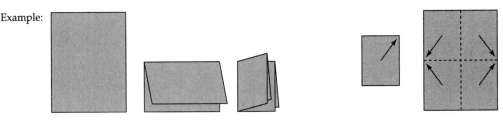

In each part, select the diagram (1, 2, or 3) that shows what the given paper will look like when it is unfolded. Briefly summarize your thinking process.

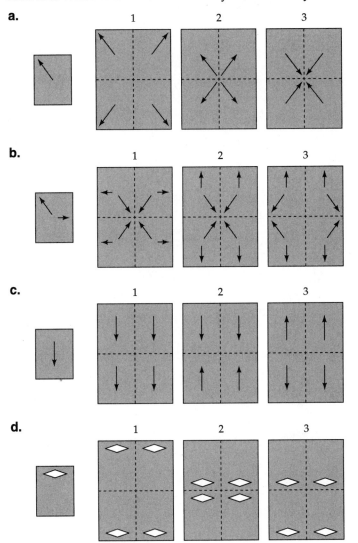

a.

b.

c.

d.

[1] "Which Way Will the Arrow Point?" from *Seeing Shapes* by Ernest R. Ranucci. *In Geometry and Visualization* by Mathematics Resource Project. © 1977. Creative Publications, Inc.

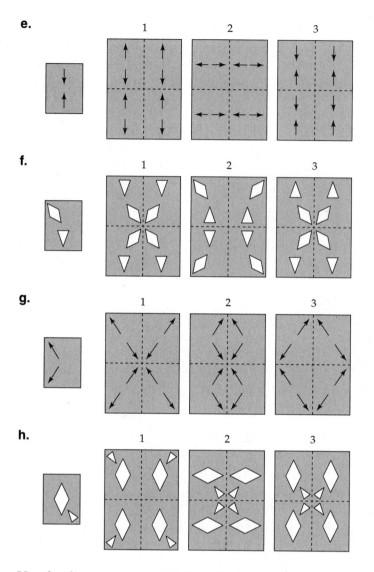

2. Use the diagrams on p. 289. In each part, a square piece of paper is folded and cut as shown in the left figure. In parts (a) through (d), the square paper is folded in half once. In parts (e) through (h), the square piece of paper is folded in half twice. The dotted lines represent the fold lines. In each case, use the right figure to draw what the design will look like when the paper is unfolded. Briefly summarize your thinking process.

Figures for   EXPLORATION 9.6,
PART 2: What happens when you unfold the paper?

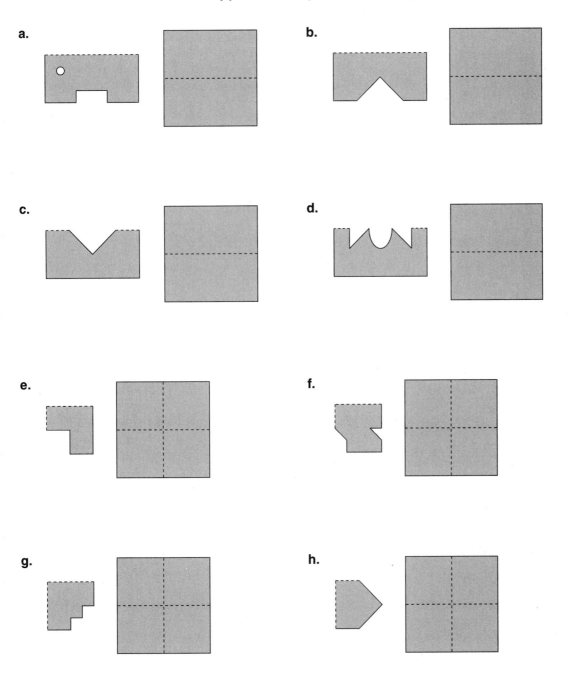

PART 3: Making copies of designs

For each of the questions in this part, whether you produce the design on your first attempt or after several attempts, turn in your solution and a summary of your thinking process. In cases of more than one attempt, include your "failures" and a description of what you learned from each "failure." (Many students say they learned a lot from their wrong answers.)

1. Fold a square piece of paper in half. Determine the cut(s) needed to produce each design when the paper is unfolded.

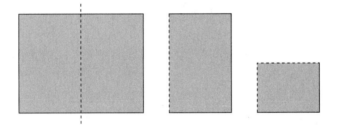

a.   b.   c.

d.   e.

2. Fold an $8\frac{1}{2}$ by 11 inch sheet of paper in half and then in half again, as shown.

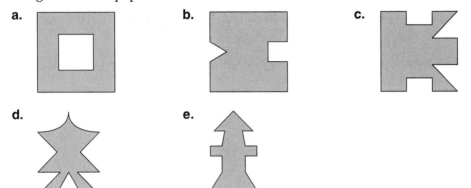

Determine the cut(s) needed to produce each design when the paper is unfolded.

a.   b.   c.

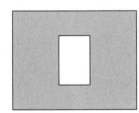   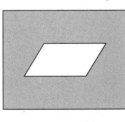

d.   e.   f.

   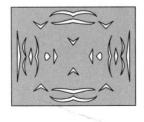   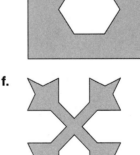

Source: Excerpted from *Symmetry: A Unifying Concept.* © 1994 by Istvan and Magdolna Hargittai. Reprinted by permission.

3.   Perform the following steps on a square piece of paper. If all goes as planned, you should end up with the paper containing, when unfolded, six 60-degree angles.

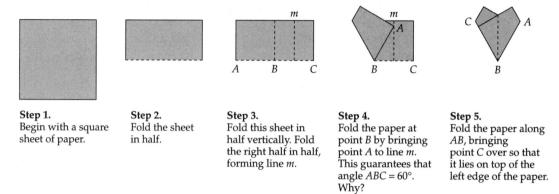

Step 1.
Begin with a square sheet of paper.

Step 2.
Fold the sheet in half.

Step 3.
Fold this sheet in half vertically. Fold the right half in half, forming line *m*.

Step 4.
Fold the paper at point *B* by bringing point *A* to line *m*. This guarantees that angle *ABC* = 60°. Why?

Step 5.
Fold the paper along *AB*, bringing point *C* over so that it lies on top of the left edge of the paper.

Describe how you would cut the folded paper in Step 5 to make each of the following.

a.   A regular hexagon   b.   An equilateral triangle   c.   A six-pointed star

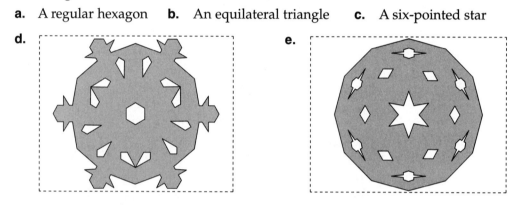

d.

e.

f.   Find a picture of a snowflake on the Internet or in a book and make a paper replica of that snowflake.

EXPLORATION 9.7   **Tessellations**

Section 9.1 includes some basic facts about tessellations—basic figures that do and don't tessellate, and why. These explorations are designed to enable you to delve more deeply into tessellations—to connect geometric ideas you have learned in Chapter 8 and Section 9.1 and to bring to life other aspects of mathematical thinking that it is important that you develop: making and testing predictions.

PART 1: Which figures tessellate?

Look at the six pairs of figures following. Work with several copies of the enlarged versions of the figures in order to determine which of the figures in each pair will tessellate: both, just one, or neither.

If both of the figures tessellate, make and justify a generalization: All figures that "look like" these two and have the following characteristics [describe the characteristics], will tessellate.

If just one of the figures in a pair tessellates, explain why the one does tessellate and why the other doesn't. That is, what characteristics or properties does the one figure have that the other doesn't have?

If neither of the figures tessellates, describe whether you believe that modifications of the figure could be made so that figures that "look like" these two might tessellate.

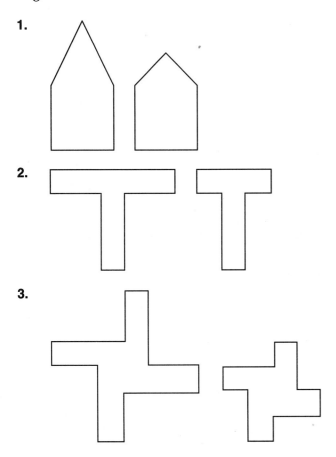

1.

2.

3.

4.

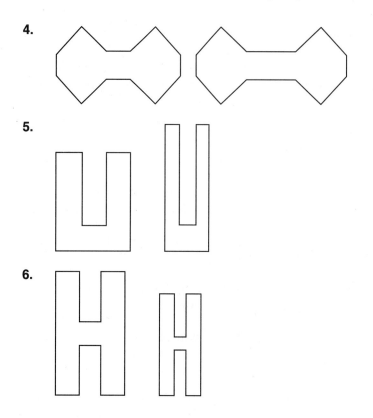

5.

6.

PART 2: What kinds of pentagons will tessellate?

As is explained in the text, the regular pentagon does not tessellate. However, some pentagons *do* tessellate. Under what conditions will a pentagon tessellate? Explore this question by making and testing several different kinds of pentagons.

   Tips: Fold a piece of paper in half, then in half again, and then in half again. Now if you draw a pentagon on the page and then cut out the pentagons (with heavy scissors), you will have eight copies of the pentagon. Alternatively, you can make a pentagon using a software program and cut and paste many copies of the pentagon.

Prepare a report that includes

- Your conclusion(s) in the following form: A pentagon with these characteristics [describe the characteristics] will tessellate. If you find more than one family that tessellates, write the description of the characteristics of each family.

- Your justification of your conclusion(s).

- A brief summary of your solution path: How did you come to your conclusion(s)? This will include your "failures" as well as your "successes."

PART 3: What kinds of arrows will tessellate?

Believe it or not, there are many kinds of arrows that will tessellate. In this exploration, your challenge is to determine the characteristics that are necessary in order for an arrow to tessellate. For example, does it need to be symmetric? Do the sides of the shaft need to be parallel? Can the tip be skinny? Make and test several different kinds of arrows, such as those shown below.

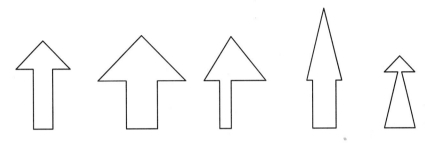

Prepare a report that includes

- Your conclusion(s) in the following form: An arrow with these characteristics [describe the characteristics] will tessellate. If you find more than one family that tessellates, write the description of the characteristics of each family.

- Your justification of your conclusion(s).

- A brief summary of your solution path: How did you come to your conclusion(s)? This will include your "failures" as well as your "successes."

PART 4: Semiregular tessellations

Now let us expand our discussion to combinations of figures that tessellate. A *semiregular tessellation* occurs when two or more regular polygons tessellate and every vertex point is congruent to every other vertex point.

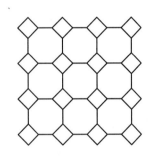

1. **a.** Using your understanding of congruent, try to define *congruent vertex point.*
   b. Compare definitions with your partner(s). Modify your definition, if needed, as a result of the discussion.

2. Cut out the figures on the Regular Polygons sheet at the end of the book. Use them to explore combinations of simple geometric figures to determine other semiregular tessellations.

   a. Sketch your "successes" and "failures" and record your reflections after each success or failure.
   b. How many semiregular tessellations do you think there are—5, 10, 20, 50, 100, thousands, an infinite number? Explain your reasoning.

3. In your explorations, you may have found some *demiregular tessellations*, which occur when two or more regular polygons tessellate but not every vertex point is congruent to every other vertex point. Show any demiregular tessellations that you have discovered. Explain why they are not semiregular tessellations.

PART 5: Escher-like tessellations

Using the language of translations, reflections, and rotations, we are now in a position to understand how the artist M. C. Escher made his tessellations (see, for example, Figure 9.21 on page 502 of the text). Escher began with a polygon that would tessellate and then transformed that polygon. We can use the analogy of chess to understand how he did this: There are legal moves and illegal moves.

1. *Translations* One legal move is to modify one side of the polygon and then translate that modification to the opposite side. For example:

   Begin with a square (Step 1).

   Modify one side (Step 2).

   Translate the modified side to the opposite side (Step 3).

   In this case, the translation is a vertical slide.

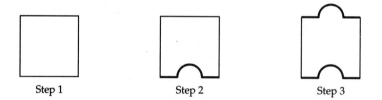

| Step 1 | Step 2 | Step 3 |

   a. Begin with a square and experiment with translations. Show your work and reasoning so that your instructor can follow your thinking. Make notes of your observations as they happen: patterns you see, conjectures you decide to pursue, questions you have.
   b. Summarize these (observations, patterns, and hypotheses about translations and tessellations.

2. *Rotations* Another legal move is to rotate part of a shape. The center of rotation—that is, the "hinge" about which the part rotates—can be a vertex or a midpoint of one of the sides. For example:

   Begin with an equilateral triangle (Step 1).

   Modify one side (Step 2).

   Rotate that side 60 degrees counterclockwise about the bottom left vertex of the triangle (Step 3).

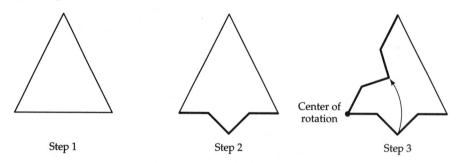

| Step 1 | Step 2 | Step 3 |

   a. Begin with a square or an equilateral triangle and experiment with rotations. Show your work so that your instructor can follow your thinking. Make notes of your observations as they happen: patterns you see, conjectures you decide to pursue, questions you have.
   b. Summarize what you learned about rotations and tessellations.

3.   *Combinations of translations, rotations, and reflections*   We can combine translations, rotations, and reflections to make tessellations. For example, we can begin with a rectangle (see Step 1 below) and, using a compass and straightedge, construct a semicircle whose diameter is half the length of the side of the rectangle (Step 2).

We can then rotate the semicircle 180 degrees, as shown in Step 3 below, the center of rotation being the midpoint of the top side of the rectangle.

Next, we translate this shape to the bottom side of the rectangle. The result is shown in Step 4. This shape will tessellate, but it's not very interesting.

However, we can also transform the shorter side of the rectangle, as shown in Step 5. This time, we replace the top half of the right side of the rectangle with a quarter-circle whose radius is equal to the length of the line segment we are replacing.

Then we *reflect* (another legal move) that quarter circle across a line that is parallel to the base of the rectangle and goes through the middle of the rectangle. See step 6.

Finally, we translate this side of the rectangle to the other side, as shown in Step 7. The original rectangle is shown with dotted lines.

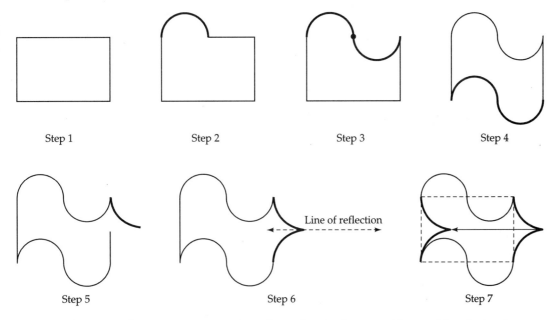

Step 1   Step 2   Step 3   Step 4

Step 5   Step 6   Line of reflection   Step 7

a.   Begin with a square or a rectangle and experiment with combinations of translations, reflections, and rotations. Show your work so that your instructor can follow your thinking.

b.   Summarize what you learned about combinations of translations, rotations, and reflections and tessellations.

4. Each of the figures below tessellates, and each began as a basic geometric figure. Can you determine the starting polygon and how the shape was made?

a.

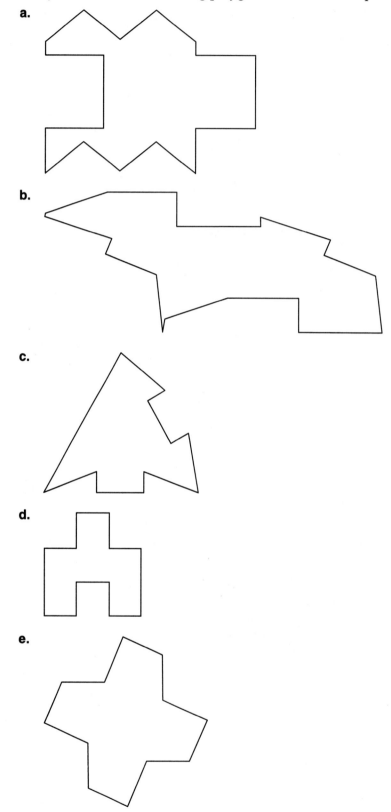

b.

c.

d.

e.

f.

g.

h.

i.

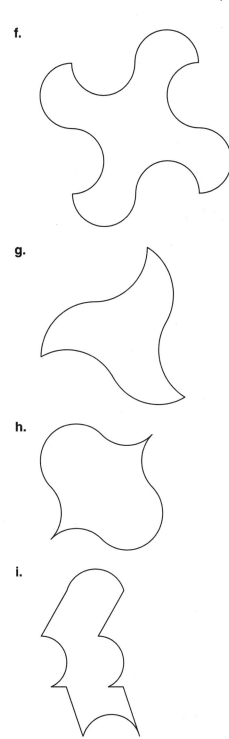

5. Make your own tessellations!

SECTION ◆ **9.2** **SYMMETRY AND TESSELLATIONS**

Symmetry is one of those mathematical concepts that we find virtually everywhere! We find it in the natural world—at the microscopic level and at the galactic level. We find it in plants and animals. We find it in everyday life and in the arts. In some cases, it has aesthetic value; in other cases, very practical value. We find it in every human culture ever known!

EXPLORATION 9.8 **Symmetries of Common Polygons**

1. Name each shape and determine its rotation and reflection symmetries. Describe any questions or problems you have.

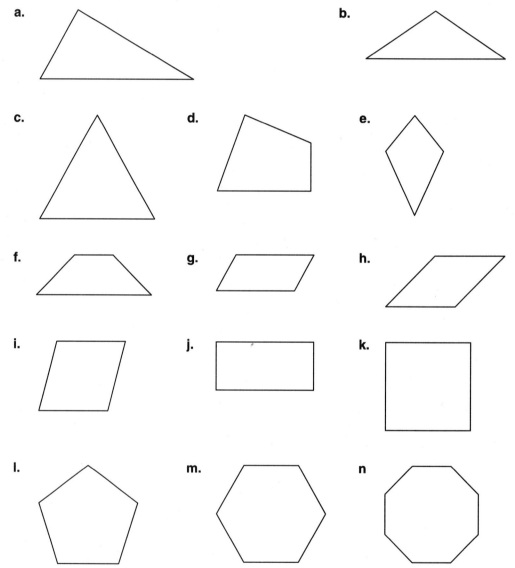

a.

b.

c.

d.

e.

f.

g.

h.

i.

j.

k.

l.

m.

n

2. Discuss how to communicate your findings from Step 1.
3. Make up and test your own figures for rotation and reflection symmetry.

EXPLORATION 9.9 **Quilts**

There are so many places to explore symmetry and so little time in this course! I have focused on quilts because quilts are rich in symmetry, they can be explored with young children, and they lend themselves to an interdisciplinary approach to learning literature, history, and mathematics.

Quilting has a fascinating and long history. Although we cannot definitely mark the date of the first quilt, we know that in the transition from hunter-gatherers to farmers many thousands of years ago, the development of agriculture inspired experiments with new ways to keep warm at night other than using animal hides, leaves, straw, and other such material. The oldest remnants of what we might call quilts—that is, spreads with patterns on them—are well over two thousand years old. Various cultures developed different patterns and techniques. We know that the Crusaders brought the idea of quilting back to Europe from the Middle East. However, an event in the fourteenth century caused quilting to come into its own: Europe suffered through what is referred to as the Great Freeze—year after year of very severe winters. This added need for more warmth at night spurred many advances in quilting, including the quilting frame. After the Great Freeze, a much greater percentage of women made quilts, and it was one of the few activities that was engaged in by poor and rich alike!

The origin of what has been called patchwork quilting is also interesting. It developed out of pioneer life in the early years of this country's history. As pioneers moved westward, more and more people lived far from stores, and the notion of "having to make do with what you've got" became a part of life. The amount of garbage generated by pioneers was a small fraction of what families throw away today. After making clothes, women would have various pieces of leftover fabric. Rather than throw them away, they would cut the pieces into small triangles, squares, and other geometric figures. The triangle was a common choice because one could get the most from scraps by cutting out triangles. (Why is this?) In many areas, wintertime was the quilting party season because there was little farm work to be done. Women would gather to trade scraps, share ideas, work together, and socialize.

Many of the quilt designs in this exploration have a rich history. Almost any public library will have many books on quilting, and with a bit of research you can probably find a quilting group in your own town!

PART 1: Beginning simple

The *block* is the basic unit of a quilt. Blocks are usually squares (and we often call them *squares*), but some quilts' blocks are hexagons or other shapes.

1.   **a.**   Examine the quilt block shown at the left below. This basic unit has been constructed from eight right isosceles triangles using two colors. Describe the construction of this block in terms of translations, rotations, and reflections. You are describing the block's *pattern,* which is sometimes called the *quilt's* pattern. (Quilting terminology is not always precise!)

     b.   Examine the shape shown at the right below, which was created by putting together four copies of the quilt block. (A quilt is made by sewing many blocks together.) Describe the new designs and shapes that you see. Explain why these designs came about.

A block   Four blocks

2.   Take eight right isosceles triangles and put them together in some way to form a block. Your block does not need to be a square. Sketch your pattern; that is, sketch what your block looks like.

3.   Work in groups of four.

     a.   Select one person's block and have each member copy that block.

     b.   Predict the shapes that will emerge when you put these four congruent blocks together.

     c.   Put the four blocks together. Describe the new patterns and shapes that you see. Explain why these patterns came about.

     d.   Repeat the process.

     e.   Sketch one figure that emerged from combining the four congruent blocks.

     f.   Describe where the new patterns came from.

PART 2: What do you see?

1.   Your instructor will show a quilt pattern. What do you see? Write your observations.

2.   Your instructor will give instructions for drawing a quilt pattern from memory. (Reminder: The objective here is not just to get it right but also to identify major structures that will enable you to generate the design.)

     a.   Draw the pattern from memory.

     b.   Alternatively, describe how you would draw the pattern from memory.

PART 3: Exploring actual quilt patterns

"There are two kinds of people in this world: those who divide everything into two groups and those who don't."[2] We hope that the various times you have been given a set of objects and asked to separate them into subsets, you have found value in the exercise. We will do this with quilts, for each quilt has many attributes.

1.   Take a minute or so for each member of your group silently to examine each quilt pattern on the next page. Then, one by one, describe what you see. That is, member 1 describes what is seen in the first pattern. If member 2 saw something different, that is noted. After each member has noted any additional observations, examine the second pattern. This time member 2 goes first, and so on.

2.   Take out the next page and cut out the quilt blocks. Separate these blocks into two or more groups in such a way that each group has one or more common attributes. Describe the common attribute(s) and name each group. Repeat this as many times as you can in the time provided. (*Note*: Save these 18 quilt blocks for use in Part 8.)

3.   What did you learn from classifying the blocks and discussing the classifications?

All quilt patterns have a story. The stories behind some of these quilt blocks are given below. These stories are adapted from *Eight Hands Round: A Patchwork Alphabet*, by Ann Paul, one of many interesting books on quilting.[3]

Old Tippecanoe:

This was the nickname of William Henry Harrison, the ninth U.S. President. When he was a general, he defeated a group of Native Americans at the Tippecanoe River in Indiana. I am not sure whether this pattern was inspired by a representation of the battle or some other characteristic of Harrison.

Storm at Sea:

This pattern represents a lighthouse, with the center square representing the light. Before modern technology, lighthouses were built at the most dangerous points along the coast.

Windmill:

Windmills used the wind as a source of power so that people could mill their grain. That is, the windmill (and water wheel) would generate enough power to turn a millstone, which would crush the grain and turn it into flour.

Yankee Puzzle:

Ann Paul notes that many years ago, people in New England often played with a puzzle that had seven small pieces—five triangles, one square, and one

---

[2]Kenneth Boulding, cited in Michael Serra, *Discovering Geometry—An Inductive Approach* (Berkeley, CA: Key Curriculum Press, 1993), p. 88.
[3]Ann Paul, *Eight Hands Round: A Patchwork Alphabet* (Harper Collins: NY, 1991).

parallelogram—which they would arrange into different designs. I find the parallels between this and the Chinese tangrams interesting. Note: The Yankee Puzzle quilt block does not contain the square and parallelogram. However, if you put several blocks together, squares and parallelograms appear!

Quilt Blocks for EXPLORATION 9.9, **PART 3**

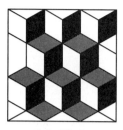

Baby Blocks

Broken Windows

Cross Roads

Does and Darts

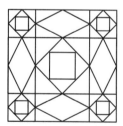

Drunkard's Path

Flower Basket

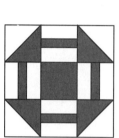

Hole in the Barn Door

Kansas Sunrise

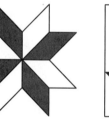

Le Moyne Star

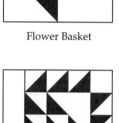

May Basket

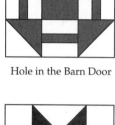

Saw Tooth Star

Star

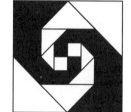

Storm at Sea

Snail's Trail

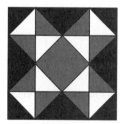

Old Tippecanoe

Tulip Basket

Windmill

Yankee Puzzle

PART 4: Closest relative

Look at the quilt blocks below. Find the "closest relative" to Indian Star. Justify your choice. There is no "right" answer. The quality of your answer depends on your justification. This task involves describing the commonalities that you find that make the two patterns so similar. Note that this is a communication task and a reasoning task. How well you do depends partly on applying problem-solving tools and making connections.

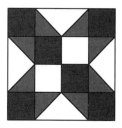

Indian Star

25-Patch Star

African Safari

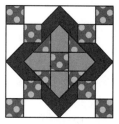

Grand Prix

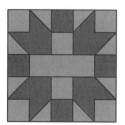

Hearth and Home

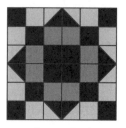

Oregon Trail

Prairies 9 Patch

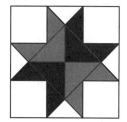

Right Hand of Friendship

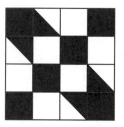

Road to Oklahoma

Weathervane

PART 5: Using transformations to construct a quilt pattern

Your instructor will ask you to construct a quilt pattern by beginning with the smallest possible piece(s) and then using slides, flips, and turns to generate the pattern. The example below illustrates one solution path for Dutch Man's Puzzle on p. 308.

Begin with the square shown in Step 1. Reflect this square through its right side (Step 2).

Translate this rectangle up one unit (Step 3).

Rotate the figure from Step 3 counterclockwise through a 1/4 turn (90 degrees); the center of rotation is the top left corner of the figure. The result is shown in Step 4.

Rotate the figure from Step 4 counterclockwise through a 1/2 turn (180 degrees); the center of rotation is the midpoint of the left side of the figure. The result is shown in Step 5.

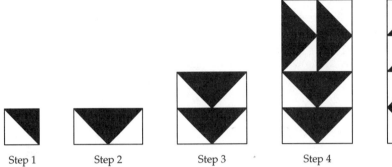

Step 1	Step 2	Step 3	Step 4	Step 5

PART 6: Similarities and differences

For each pair of quilt patterns below, describe how the two patterns are alike and how they are different.

1.

Star Puzzle   Fourth of July

2.

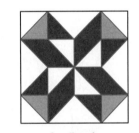 

Rocky Road   Railroad

3.

Dutch Man's Puzzle   Mosaic

PART 7: Quilts and tessellations

Below are a variety of shapes taken from quilt blocks. In each case, predict whether the shape will tessellate.

a.   Describe your reasoning, to the extent that you can. Recall the van Hiele discussion in the text (we want you to move beyond level 1), and recall the importance of making connections. Can you apply what you already know about tessellations, about transformations, about symmetry, and/or about geometry (the sum of the angles of a quadrilateral is 360 degrees, breaking apart and putting together figures, etc.)?

b.   Determine whether the figure tessellates—show your work. You can do this by tracing, by cutting out figures, or by making more copies.

c.   If your prediction was wrong, try to explain why the figure either does or doesn't tessellate. If your prediction was right, but you saw new things, describe those new insights.

d.   If the figure doesn't tessellate, can you modify it so that it will?

1.

Nelson's Victory

2.

Windmill

3.

T

4.

Yankee Puzzle

5.

Broad Arrow

6.

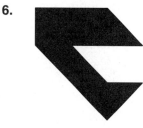

Belt Buckle

7.

Broken Sugar Bowl

PART 8: Different symmetries

Take out page 311 and cut apart the 12 quilt patterns. Combine these patterns with the set of quilt patterns in Part 3.

1. Determine which patterns have rotation symmetry. Classify them according to the kind of rotation symmetry that they have.
2. Determine which patterns have reflection symmetry. Classify them according to the kind of reflection symmetry that they have.

Here are the stories behind several of the quilt patterns on page 311.

Anvil:

In earlier times, most towns had a blacksmith who made things from iron by placing the metal in a fire until it became soft enough to have its shape changed by pounding it with a hammer against an anvil. The blacksmith made axes, farm tools, horseshoes, and other things that had iron components. The anvil had both a straight part and a curved part. When the blacksmith wanted to make the iron straight, he placed the iron on the flat part of the anvil. When he wanted to make the iron curve, he placed it on the curved part of the anvil.

Eight Hands Round:

At various times, people in an area would gather together: for a barn raising, a quilting bee, or some other occasion. In the evening, there would usually be some music and some dancing—square dancing and/or contra dancing. Someone would "call" the dances—that is, give instructions throughout the dance. One of the calls was "eight hands round," which meant for the four couples in a square to join hands in a circle.

Log Cabin:

When the American settlers moved to the frontier, they had to make their own homes from the wood in the forest. You have surely heard the stories about Abraham Lincoln's being born in a log cabin. This pattern was inspired by someone's creative representation of the pattern of the logs in a log cabin.

Underground Railroad:

The Underground Railroad was a network of people who helped slaves escape from slavery. A slave trying to get to a state where slavery was illegal often had to travel hundreds of miles. Those people who helped the slaves fed and hid them, usually at night, and then gave them directions to the next safe place.

3. Using different colors of construction paper, cut out a number of squares and triangles.
    a. Make a pattern (a block) that has rotation symmetry but not reflection symmetry. Briefly convince the reader that the figure does have rotation but not reflection symmetry.
    b. Make a pattern that has reflection symmetry but not rotation symmetry. Briefly convince the reader that the figure does have reflection but not rotation symmetry.
    c. Make a pattern that has both reflection symmetry and rotation symmetry. Briefly convince the reader that the figure does have rotation and reflection symmetry.

Quilt Patterns for EXPLORATION 9.9, PART 8

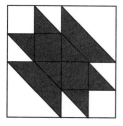

Anvil

Crazy Ann

Dutch Man's Puzzle

Eight Hands Round

Flying Geese

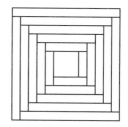

Log Cabin

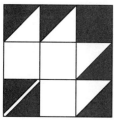

Maple

Peony

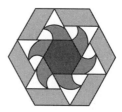

Starry Nights

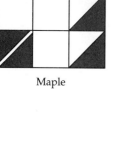

Susannah

Rabbit Paw

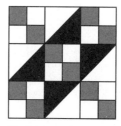

Underground Railroad

EXPLORATION 9.10 **Symmetry Groups**

In this section, we have seen that there are many connections between operations on shapes and operations on numbers. This exploration seeks to deepen that understanding. Let us first review what we know about one operation on one set of numbers: addition on the set of integers.

- We found that the operation of addition is associative; that is, $(a + b) + c = a + (b + c)$.

- We found that there is an identity (0) for that operation.

- We found that there is an inverse for every number.

When a set (in this case, a set of numbers) and an operation satisfy these three criteria, we call the relationship a group. To have a set and an operation that have this relationship implies a certain amount of structure. Groups are critically important in understanding the underlying structure of many mathematical and real-world phenomena.

Many operations with many sets have this same underlying structure. The sets can be numbers, they can be shapes, and they can be many other things. The operations can be addition, they can be reflection, and they can be many other things. However, groups appear all over the landscape of mathematics, and the analysis of groups has led to discoveries in science and other fields.

We see groups in geometry too. Let us explore what is called the symmetry group of the equilateral triangle. In this case, our set is the set of symmetries of an equilateral triangle, and our operation is the composition of those transformations. Now this is a mouthful for most students, so let me put in another way: Mathematicians have found that the collection of the symmetries of any figure will always be a group. We will explore this with "interesting" figures: equaliteral triangles and squares.

As you have discovered, the equilateral triangle is the triangle with the most symmetry; it is also a regular polygon. It has threefold rotation symmetry and three lines of reflection symmetry. Let us examine the combination of these symmetries. For this, you will need to cut out the six equilateral triangles on page 317.

PART 1: Equilateral triangles

1. **a.** Cut out the six triangles on page 317. Label the first triangle as shown in the bottom left triangle. Then label the other triangles appropriately. That is, where will vertices 1, 2, and 3 be after the specified turn or reflection? *Note:* In each of the five cases, you always begin with the triangle in the original position. We will use the symbols I, r, r_2, m_1, m_2, and m_3 to represent these six symmetries of the triangle — that is, each of these transformations results in the triangle lying on top of itself.

Original position	After 1/3 turn clockwise	After 2/3 turn clockwise	After vertical reflection	After this diagonal reflection	After this diagonal reflection
I	r	r_2	m_1	m_2	m_3

b. Fill in the "multiplication" table that appears below. In each case, you begin in the original position, so the top left cell represents the identity operation—that is, do nothing. This is like 1×1 in multiplication.

Look at r times r. That is, if you perform a 1/3 turn and then another 1/3 turn, the triangle will be in the same position as if you had done a 2/3 turn. Thus, r times $r = r_2$.

One more example: Do a 1/3 turn and then do reflection m_1—that is, vertical line. What do you get? If you did it correctly, you found that this composition is equal to doing the m_3 reflection to the original figure.

$\triangle$	I	r	r_2	m_1	m_2	m_3
I	I					
r		r_2		m_3		
r_2						
m_1						
m_2						
m_3						

c. Compare results with a partner to make sure you are accurate.

d. What do you see? (This is an intentionally wide-open question).

e. What patterns do you see? Write down all the patterns you see in the table. This may take some time, so you need not write in full, grammatically perfect sentences. Just record enough so that you will be able to refer to these notes during the class discussion.

f. Select one nontrivial pattern to describe carefully. Exchange descriptions with a partner. If your partner was able to understand the pattern, great. If not, circle the words or phrases that caused the problem. Then rewrite and exchange with another partner.

g. Do this set and this operation satisfy the definition of a group?

(1) Is there an associative property $(a \times b) \times c = a \times (b \times c)$? What is your support for your answer?
We know from before that I is the identity operation for this set.

(2) Does each operation have an inverse? If so, write the inverse of each operation.

$I^{-1} = $ ____ $r^{-1} = $ ____ $r_2^{-1} = $ ____ $m_1^{-1} = $ ____ $m_2^{-1} = $ ____ $m_3^{-1} = $ ____

2. Careful examination enables us to realize that we don't *need* three different mirrors. That is, the m_2 reflection is equivalent to the vertical reflection and a 1/3 turn. Thus the m_2 reflection is equivalent to the m_1 and r composition. Similarly, the m_3 reflection is equivalent to the vertical reflection and a 2/3 turn. This discovery simplifies our notation, because we need only specify one line of reflection: vertical. And so the new multiplication table for the symmetries of an equilateral triangle looks like the table below. That is, analysis of our operations enables us to change our notation to reflect the relationships/connections that we found. Fill in the table below.

$\triangle$	I	r	r_2	m	mr	mr_2
I						
r						
r_2						
m						
mr						
mr_2						

PART 2: Squares

For this part, you will need to cut out the eight squares on page 317.

1. **a.** Label the first square as shown to the right. Then label the other squares appropriately. That is, where will vertices 1, 2, 3, and 4 be after the specified turn or reflection?

1	2
4	3

   b. Fill in the table below to determine the multiplication table for the symmetries of a square.

$\square$	I	r	r_2	r_3	m	mr	mr_2	mr_3
I								
r								
r_2								
r_3								
m								
mr								
mr_2								
mr_3								

   c. Compare results with a partner to make sure you are accurate.
   d. What do you see?
   e. What patterns do you see?

f. Select one nontrivial pattern to describe. Exchange descriptions with a partner. If your partner was able to understand the pattern, great. If not, circle the words or phrases that caused the problem. Then rewrite and exchange with another partner.

g. Do this set and this operation satisfy the definition of a group?

   (1) Is there an associative property $(a \times b) \times c = a \times (b \times c)$? What is your support for your answer?
   We know from before that I is the identity operation for this set.

   (2) Does each operation have an inverse? If so, write the inverse of each operation.

Figures for EXPLORATION 9.10, **PART 1 and PART 2**

PART 1: Equilateral Triangles

PART 2: Squares

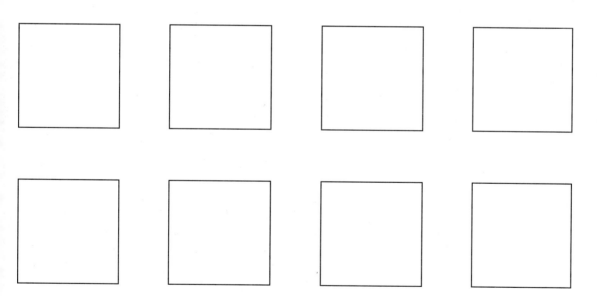

SECTION   **9.3**   **EXPLORING SIMILARITY**

EXPLORATION 9.11   **Similarity with Pattern Blocks**

In Chapter 5, we used pattern blocks to explore fraction concepts. These blocks also let us see geometry "in action."

PART 1: Triangles

1. Take the small green triangle. Using the pattern blocks, make a bigger triangle that is similar in shape to this triangle. Explain why you believe this triangle is similar to the original triangle.
2. Make another triangle that is similar to the original triangle. Explain why you believe this triangle is similar to the original triangle.
3. Record other observations: patterns, conjectures, and questions. For example, are there patterns in the lengths of the sides? In how large the figures are? In how you actually make larger figures?

PART 2: Squares

1. Take the small orange square. Using the pattern blocks, make a bigger square that is similar in shape to this square. Explain why you believe this square is similar to the original square.
2. Make another square that is similar to the original square. Explain why you believe this square is similar to the original square.
3. Record other observations: patterns, conjectures, and questions. For example, are there patterns in the lengths of the sides? In how large the figures are? In how you actually make larger figures?

PART 3: Parallelograms

As it turns out, all equilateral triangles are similar to one another, and all squares are similar to one another. This is not true for all parallelograms!

1. Take the small blue parallelogram. Using the pattern blocks, make a bigger parallelogram that is similar in shape to this parallelogram. Explain why you believe this parallelogram is similar to the original parallelogram.
2. Make another parallelogram that is similar to the original parallelogram. Explain why you believe this parallelogram is similar to the original parallelogram.
3. **a.** Write your first draft of a definition of *similar* with respect to parallelograms. Your definition should be more precise than "same shape, possibly different size."
    b. Compare your definition with that of your partner(s). After the discussion, modify your definition, if needed.
4. Record other observations: patterns, conjectures, and questions. For example, are there patterns in the lengths of the sides? In how large the figures are? In how you actually make larger figures, and so on?

PART 4: Trapezoids

1. Take the small red trapezoid. Using the pattern blocks, make a bigger trapezoid that is similar in shape to this trapezoid. Explain why you believe this trapezoid is similar to the original trapezoid.

2. Make another trapezoid that is similar to the original trapezoid. Explain why you believe this trapezoid is similar to the original trapezoid.

3. Record other observations: patterns, conjectures, and questions. For example, are there patterns in the lengths of the sides? In how large the figures are? In how you actually make larger figures, and so on?

4. Which of the following trapezoids is similar to the small red trapezoid? Justify your reasoning.

5. **a.** Write your first draft of a definition of *similar* with respect to any geometric figure. Be precise, as noted in Part 3.

   b. Compare your definition with that of your partner(s). After the discussion, modify your definition, if needed.

Extension

There is a pattern concerning the area of consecutive similar pattern blocks that is true for all of the blocks: The ratio of the area of the next bigger similar figure to the original is 4:1. That is, if we count the area of the green triangle as 1 unit, the next bigger equilateral triangle has an area of 4 units. The blue parallelogram has an area of 2 units, and the next bigger blue parallelogram has an area of 8 units. Thus, the ratio of their areas is 4:1. This is also true for the orange square, the red trapezoid, and the yellow hexagon.

Do you think this is true only in these cases, or will it be true for composite figures too? For example, will the area of the next bigger figure similar to this one be 4 times as great as the area of this one?

EXPLORATION 9.12   **Similar Figures**

PART 1: Similar triangles

In high school geometry, we discovered that we did not have to show that all six pairs of corresponding angles and sides were congruent in order to know that two triangles were congruent. We learned about SSS, SAS, ASA, AAS, and HL. What might we need to show in order to know that two triangles are similar? Once again, the goal is to find shortcuts. At this point, in order to know that the two triangles are similar, we need to measure the lengths of all six sides and all six angles, and then we have to determine three ratios. If all the ratios are equal and all three pairs of angles are equal, then we know the two triangles are similar.

$$\frac{AB}{PQ} = \frac{AC}{PR} = \frac{BC}{QR} \quad \text{and} \quad m\angle A = m\angle P, m\angle B = m\angle Q, m\angle C = m\angle R$$

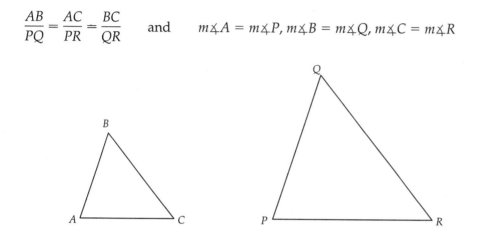

The question is: Do we have to know all the ratios to determine whether two triangles are similar? What combinations of sides and angles will be sufficient?

1.   Write your initial hypothesis and reasoning about *one* combination that is sufficient to prove that two triangles are similar.
2.   Discuss your ideas with your partner(s). Take one hypothesis; try to convince yourself that it is true, and try to make triangles that are not similar but that satisfy your hypothesis.
3.   Present your hypotheses to another group. Note any comments or suggestions made by the other group.
4.   Work through Part 1 repeatedly (as time permits) in order to convince yourself of as many similarity combinations as you can.

PART 2: Similar quadrilaterals

The figure below shows a pair of similar rectangles and a pair of similar trapezoids. At this stage of our explorations, to confirm that they are indeed similar, we would have to measure all eight sides and all eight angles and verify that the lengths of corresponding pairs of sides had the same ratio and that corresponding angles were congruent. The question is: Do we have to know all eight pieces of information? What combinations of sides and angles will be sufficient?

1. *Rectangles*

    a. Write your initial hypotheses and reasoning for rectangles.
    b. Discuss your ideas with your partner(s). Take one of your hypotheses; try to convince yourself that it is true, and try to make rectangles that are not similar but that satisfy your hypothesis.
    c. Present your hypotheses to another group. Note any comments or suggestions made by the other group.

2. *Trapezoids* Repeat Step 1 for trapezoids.

Extensions

Come up with a conjecture that will work for all quadrilaterals. That is, if these conditions are met, then the two quadrilaterals will be similar. Justify your conjecture.

EXPLORATION 9.13   **Reptiles**

Solomon Golumb (see Exploration 8.3) invented reptiles, figures that both tessellate and also replicate themselves (as explained in Part 1 below). What began as an interesting thought turned into an idea with powerful real-life applications.

PART 1: A trapezoid reptile

Consider the trapezoid below.

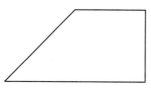

This trapezoid tessellates. Now we will see how it "replicates itself." We can divide this trapezoid into four smaller congruent trapezoids, each of which is similar to the original—sort of like having children!

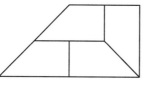

1.   Why is "reptile" an appropriate name for such figures?

2.   Prove that the four little trapezoids are similar to the larger trapezoid.

3.   **a.** Make another trapezoid that is a reptile.
         b. Describe the attributes that a trapezoid must have in order to be a reptile.

4.   How many different ways can you find in which this trapezoid will tile the plane?

5.   Some tessellation patterns have a *lattice structure* and some do not. A lattice is an arrangement of dots that follow a regular geometric pattern. The different kinds of dot paper in the cutouts are examples of lattices. A geoboard represents a simple lattice. When considering whether a tessellating polygon has a lattice structure, we need to focus on the dots that are the vertices of the polygon. Do some or all of the trapezoid reptile tessellations have no lattice structure? "Tilings without lattices are of great interest today among mathematicians and solid-state scientists because they share many strange properties with some recently discovered crystalline materials called quasicrystals."[4]

[4]Marjorie Senechal, "Shape," in *On the Shoulders of Giants*, p. 159. Reprinted with permission from *On the Shoulders of Giants: New Approaches to Numeracy.* Copyright © 1990 by the National Academy of Sciences. Courtesy of the National Academy Press, Washington, D.C.

PART 2: A hexagon reptile

The hexagon below is another reptile.

1.   Determine how to divide this hexagon into several little congruent hexagons, all of which are similar to the original.
2.   How many different ways can you find in which this hexagon will tile the plane?
3.   Do some or all of the hexagon reptile tessellations have no lattice structure?

Extensions

1.   On the basis of these two examples, hypothesize some of the characteristics that a polygon must have in order to be a reptile.
2.   Discover another polygon that is a reptile.

10

Geometry as Measurement

Many, if not most, uses of basic mathematics involve measurement. Formulas are important because they enable us to answer the "how much?" questions more efficiently. However, for many prolems, there is no formula (for example, what is the area of your state?). In other cases, you need to apply measurement ideas to solve nonroutine problems (for example, to find the height of a tree). In this chapter, you will explore important measurement ideas and also come to understand why various formulas work.

SECTION **10.1** **EXPLORING SYSTEMS OF MEASUREMENT**

As you have already realized, much of our use of numbers involves measurement: how much, how many, how far, and so on. In this section, we will examine some of the issues surrounding linear measurement—that is, situations in which we want to ask questions or gain information about distances, heights, and thicknesses. We will also work with measuring weight and begin thinking about area and volume.

EXPLORATION 10.1 **How Long?**

As you did in Alphabitia, for a moment imagine that you are back in time before there were feet or meters or standardized tools to measure them. Your tribe wants to make a community building, and they want the length and width of this building to be exactly the length and width of your classroom.

1. Using only the tools your instructor provides and your own reasoning, with your partners, determine the length and width of the room.
2. Present your method and your answer to the class.
3. What did you learn about measurement from this exploration?

EXPLORATION 10.2   **How Tall?**

In many cases, we obtain a linear measurement without measuring the object directly. For example, we might read that the distance from New York to Los Angeles is 2825 miles and that the Sears Tower in Chicago is 1454 feet tall. These numbers were not determined using tape measures! What do you do when you cannot measure something directly?

1.   Your instructor will select an object (tree or building) for which you will determine the height.
  a. Brainstorm reasons why someone might actually want to know the height of this object.
  b. Brainstorm ideas for determining the height.
2.   Select and pursue one method with your group. Describe and justify each step in your plan.
3.   a. Justify the precision in your answer—that is, the choice of unit and the degree of accuracy: for example, to the nearest 10 feet (meters), 1 foot (meter), one decimal place, and so on.
  b. Describe and explain your degree of confidence in your result.
  c. Describe any difficulties you had and how you overcame them.
4.   Present your group's findings to the class.
5.   Which group's method do you think will produce the answer closest to the actual height? Justify your response.

EXPLORATION 10.3 **How Thick?**

Most of our linear measurements use units like inches, feet, or miles, or the corresponding metric units. However, people in many occupations need measurements of amounts that are very small. For example, how thick is your skin? What is the wavelength of red light? How long is an amoeba?

Let us explore how we might answer one such question: How thick is one sheet of paper?

1. With your partner(s), brainstorm ideas for answering this question.
2. Select and pursue one method with your group. Describe and justify each step in your plan.
3. a. Justify the precision in your answer — that is, the choice of unit and the degree of accuracy, for example.
   b. Describe and explain your degree of confidence in your result.
   c. Describe any difficulties you had and how you overcame them.
4. Present your group's findings to the class.
5. Which group's method do you think will produce the answer closest to the actual thickness? Justify your response.
6. What did you learn about measurement from this exploration?

EXPLORATION 10.4   **How Much Is a Million?**

Researchers have told us that many people's behavior with numbers is very different when the size of the numbers becomes greater than they can actually imagine. For example, we can concretely imagine how much space would be taken up by 100 people listening to a musician play, but it is hard to imagine how much space would be taken up by a concert audience of 1 million people.

Many elementary school teachers have told me that when they ask, for example, "How large does a container need to be to hold one million paperclips?", children invariably underestimate. Many give responses such as "a jewelry box" or "a cereal box."

PART 1: 1 million dollars

Betty has just won 1 million dollars in 1-dollar bills!

1.   How much do you think 1 million dollar bills will weigh?
   a.   Write down your estimate and any reasoning behind the estimate.
   b.   With your partners, devise a plan for answering this question and then carry out that plan.
   c.   Present your results to the class.

2.   What if we made a road with these dollar bills, in which each row consisted of ten 1-dollar bills? How long would the road be?

   a.   Write down your estimate and any reasoning behind the estimate.
   b.   With your partners, devise a plan for answering this question and then carry out that plan.
   c.   Present your results to the class.

3.   What if we were to stack these dollar bills in a room, putting a rubber band around sets of 100 1-dollar bills? What would be the dimensions of a container that would hold one million dollar bills?
   a.   Write down your estimate and any reasoning behind the estimate.
   b.   With your partners, devise a plan for answering this question and then carry out that plan.
   c.   Present your results to the class.

PART 2: 1 million pennies

1.   How much do you think 1 million pennies would weigh?
   a.   Write down your estimate and any reasoning behind the estimate.
   b.   With your partners, devise a plan for answering this question and then carry out that plan.
   c.   Present your results to the class.

2. In *The Wizard of Oz*, Dorothy walked down the yellow brick road. What if we made a road that was 2 feet wide with 1 million pennies? How long would the road be?

    a. Write down your estimate and any reasoning behind the estimate.

    b. With your partners, devise a plan for answering this question and then carry out that plan.

    c. Present your results to the class.

3. What if we were to stack these 1 million pennies in a room, after first putting groups of 50 pennies into rolls? What would be the dimensions of a room that would hold 1 million pennies?

    a. Write down your estimate and any reasoning behind the estimate.

    b. With your partners, devise a plan for answering this question and then carry out that plan.

    c. Present your results to the class.

SECTION **10.2** **EXPLORING PERIMETER AND AREA**

We encounter the need to know perimeters and areas in our everyday and work lives. In determining perimeters and areas, we need many ideas and formulas that have been developed over the centuries.

EXPLORATION 10.5 **Tangrams and Measurement**

PART 1: Angles

1. Determine the sizes of the angles in each of the seven tangram pieces found at the end of this book *without the use of a protractor* (that is, by reasoning alone). Explain how you did this.
2. Check your answers with a protractor. If you made any mistakes, determine where you went wrong.

PART 2: Area

1. Determine the relative size of each tangram piece.

   a. First, write down what you understand the task to mean. That is, in your own words, describe what you understand "determine the relative size of each piece" to mean.
   b. Record the order in which you determined the relative sizes. That is, which did you determine first: the biggest triangle, the smallest triangle, the square, or some other piece?
   c. Now meet with a partner or with another group. Each person or group reports its findings. The other person or group listens actively. Note anything you learned from this discussion.

2. In the standard tangram package, the hypotenuse of the small triangle is 2 inches long.

   a. Determine the length of each side of each of the seven pieces without measuring, by examining relationships between pieces and using the Pythagorean theorem. (For a right triangle with legs a and b and hypotenuse c, $a^2 + b^2 = c^2$.) Explain your work.
   b. Now determine the actual area (in square inches) of each figure. Explain your work.

3. a. Why do you think the standard tangrams are the size that they are?
   b. What if the sides of the small squares were 1 inch? What would be the dimensions of the other pieces?

4. Using the dimensions of the actual tangram pieces, determine the area of the *arrow*. The shaded region is made from the tangram pieces.

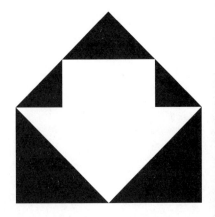

EXPLORATION 10.6   **Exploring the Meaning of Area**

What Does Area Mean?

1. Let us begin by exploring what area means—not how we determine it, but what it means. Imagine that someone from another planet came to visit, and that when you made a comment like, "The area of this figure is greater than the area of that figure," the person asked you what *area* means. How would you respond? Write your response and then compare responses with your partner(s).

What Does the Number Mean?

2. Consider a rectangle whose dimensions are 12 inches by 24 inches. We might say that the area is 288 square inches or, if we used feet as our unit of measurement, we might say that the area is 2 square feet. If we were in a country using the metric system of measurement, we would determine the length and width to be approximately 60 centimeters and 30 centimeters, and we would say that the area is 1800 square centimeters. From this perspective, there is not a functional relationship between an object and the number used to denote its area, unless you specify the unit measurement.

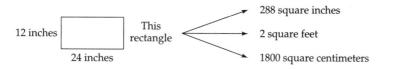

Thus we need to look at the numbers and what the numbers mean. For example, suppose you have heard that the floor space of the student center is 10,000 square feet. What does that number mean? Write your thoughts and then share them with your partners.

3. Now let us explore the meaning of numbers in measurement. Mikala has a problem:

> The area of her lawn is 1200 square feet. She wants to spread fertilizer over the lawn, but the bag of fertilizer says that one bag will cover 100 square yards. How many bags of fertilizer should she buy?

a. Think about this problem—what you know about feet and yards, and the tools you have for problem-solving (such as Polya's four steps on the inside front cover of the Explorations Manual). Write down your initial thoughts and hypotheses, including an estimate if you feel you can make one.

b. Meet with your partner(s). Select one or more ideas to pursue and then pursue them.

4. After the class discussion, write your second-draft response to the question asked in Step 2. When you see a measurement (for example, 20,000 square feet), what does that number mean?

EXPLORATION 10.7   **Exploring Area on Geoboards**

Let us explore the concept of area and some area formulas using geoboards.

PART 1: Area on the geoboard

In our explorations on the geoboard and on dot paper, we will discuss the areas of figures in reference to the unit square — that is, the area enclosed by the smallest square you can make on your geoboard.

1. **a.** Make as many "different" squares as you can on a 5 × 5 geoboard or dot paper supplied at the end of the book.
    b. Compare your results and your strategies with your partner(s).
    c. Determine the area of each of the squares.
    d. Challenge: Can you make a square with an area of 1 unit? 2 units? 3 units, and so on? Which squares can't be made? Can you explain why they are impossible?

2. **a.** Make as many "different" rectangles as you can on a 5 × 5 geoboard or dot paper provided by your instructor.
    b. Compare your results and your strategies with your partner(s).
    c. Determine the area of each of the rectangles.
    d. Challenge: Can you make a rectangle with an area of 1 unit? 2 units? 3 units, and so on? Which rectangles can't be made? Can you explain why they are impossible?

3. **a.** Determine the area enclosed by each of the figures following. Briefly describe your solution path so that a reader could see how you determined the area.
    b. Compare your answers and your strategies with your partner(s).
    c. With your partner, make an unusual shape on the geoboard. Determine the area independently, and then compare answers and solution strategies. Do this for several figures.
    d. Summarize and justify useful strategies that you developed. Imagine writing this and sending it to a friend in order to share with your friend your insights.

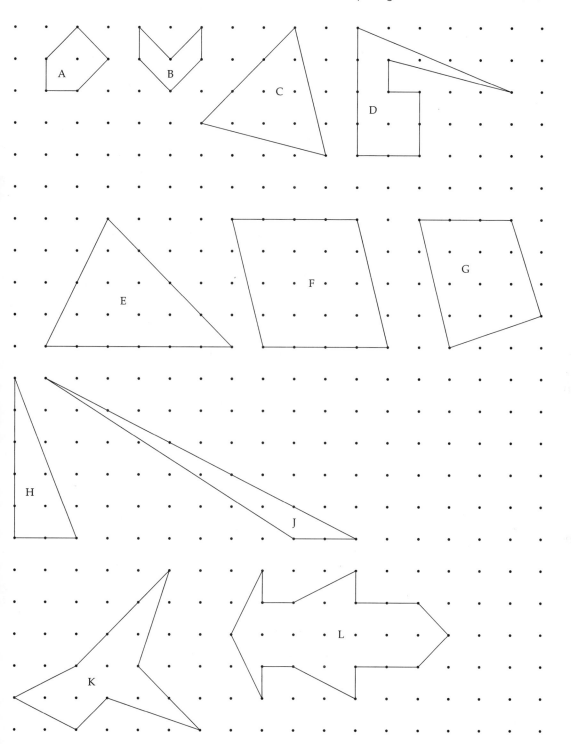

PART 2: Understanding area formulas

1. *Determining the area of parallelograms*   Let us now explore certain geometric figures and try to understand why their area formulas work. Even if you already remember certain area formulas (from previous experience), the purpose of these explorations is not to get the formula, but to understand why it works. The *why* enables us to apply our understanding to new situations. For example, if you really understand why the formulas for the area of a rectangle, a parallelogram, and a triangle work, then you can use that knowledge to construct the formula for the area of a trapezoid. In that spirit, examine the parallelograms below.

    a. Determine the area of the parallelograms below by a means other than using a formula. Can you see any patterns that hold for all four parallelograms? (Make more parallelograms if you wish.)

    b. State and justify a formula for determining the area of any parallelogram.

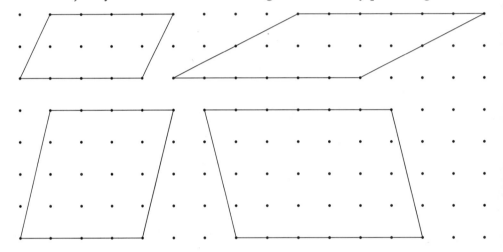

2. *Determining the area of triangles*   We can apply our understanding of the area of parallelograms to determine a way to find the area of any triangle.

    a. Determine the area of the triangles below by a means other than using a formula. Can you discern any commonalities that enable us to determine the area of any triangle? (Make more triangles if you wish.)

    b. State and justify a formula for determining the area of any triangle.

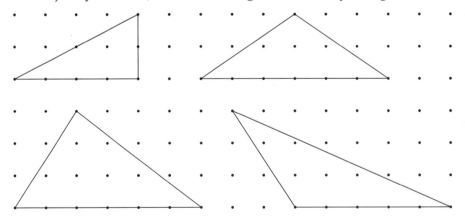

3.   There is an old saying that "seeing is believing." Determine the areas of the triangles below — the right triangle, the obtuse triangle, and the isosceles triangle. What did you discover? Can you explain why what you discovered is true?

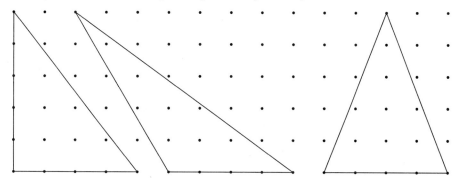

4.   *Determining the area of trapezoids*

   a.   On your geoboard or dot paper (supplied at the end of the book), make several trapezoids, and determine the area of each trapezoid.
   b.   Gather data on those attributes on which the area seems to depend. Can you see patterns in the numbers, and/or can you apply your work in the previous steps in this exploration so that you can draw a new trapezoid and predict the area on the basis of certain measurements?
   c.   After you are finished, write a chronological report. That is, tell the reader not only what you discovered but also how you discovered it. If you went down blind alleys or made and then discarded a hypothesis, help the reader to see why the blind alley or hypothesis made sense at the time and what you learned from that work.

EXPLORATION 10.8

**Exploring Relationships Between
Perimeter and Area**

The relationship between perimeter and area is a rich field for exploration because there are many possible ways in which these two attributes can be related.

PART 1: Perimeters and areas on geoboards

1. **a.** Suppose someone doubled the area of her garden. Does that mean that the length of the fence around the perimeter of her garden doubled also? What do you think?
   b. If we know that the area of one garden is 50 percent greater than the area of another garden, does that mean that the perimeter will be 50 percent greater? What do you think?
   c. Take some time to describe your present thoughts on the relationship between perimeter and area from this perspective.

The relationship between area and perimeter is not a simple one, and we are going to investigate the relationship systematically, using a technique developed by mathematicians that helps us when relationships are complicated. We are going to keep one variable constant and look at what happens to the other variable.

2. **a.** On dot paper, make a number of polygons, each of which has an area of 15 units and in each of which all sides are either horizontal or vertical line segments. Determine the perimeter of each figure. In this case, we are holding area and certain attributes of shape constant and looking at how the perimeter changes.
   b. Look at those figures with the smallest perimeters and those with the greatest perimeters. Describe differences between the figures with smaller perimeters and the figures with greater perimeters.

3. **a.** Now make a number of polygons, each of which has a perimeter of 24 units and in each of which all sides are either horizontal or vertical line segments. Determine the area of each figure. In this case, we are holding perimeter and certain attributes of shape constant and looking at how the area changes.
   b. Look at those figures that have the smallest areas and those figures that have the largest areas. Describe differences between the figures with smaller areas and the figures with larger areas.

4. Now, consider the original question: How are perimeter and area related? What do you believe now? Describe your present beliefs. If they are different from or more refined than your previous beliefs, describe the experiences, observations, or conversations that changed your beliefs.

PART 2: Changing dimensions

Let us explore the relationship between perimeter and area from another perspective.

1. What if you doubled the length of a rectangle but didn't change the width? The area would double, but what about the perimeter? What would be the effect on the perimeter? Explore this question with rectangles of different dimensions. Can you arrive at a statement that will be true for *all* rectangles? Write a report containing the following material:

    a. A brief summary of what you did.
    b. Your present description, in words and/or a formula, of the relationship between the original and the new perimeter.

2. Exchange descriptions with a partner. Provide feedback with respect to accuracy and clarity.

Looking Back on Exploration 10.8

What did you learn from this exploration?

EXPLORATION 10.9 **What Does π Mean?**

1. What does π mean? Suppose someone asked you this question. The person knows that the value of π is approximately 3.14 but does not have a sense of what π means. How would you answer that question? Write your first-draft thoughts before reading on.

2. After observing your instructor's demonstration or doing explorations provided by your instructor, describe what π means (second draft).

3. In groups of three, go through the following process:

   a. Each person reads her or his response. The others give feedback with respect to the accuracy and clarity of the response, in that order.

      ■ Accuracy: If a group member feels that the statement is not entirely accurate, discuss this issue until it is resolved.

      ■ Clarity: If a group member feels that certain words or phrases are ambiguous or unclear or vague, discuss those issues until they are resolved.

   b. Write down your third draft of "what π means."

Looking Back on Exploration 10.9

Describe any questions you have about π at this point.

EXPLORATION 10.10   **Exploring the Area of a Circle**

The purpose of this exploration is to develop an understanding of the formula for the area of a circle.

1.   Cut the circle (at the end of this book) into sectors as shown below. Arrange these sectors into a "parallelogram," as shown.

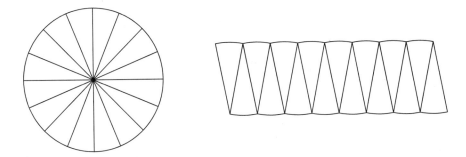

2.   If the radius of the circle is r, determine the length and width of the parallelogram in terms of r.

3.   Use this information to determine the area of the circle. Explain both the *what* and the *why* of your work.

Looking Back on Exploration 10.10

Describe any questions you have about π or circles at this point.

EXPLORATION 10.11   **Can You Make the Quilt Pattern?**

In Chapter 9 we explored quilt structures and symmetries. In this chapter, we will explore a problem that real quilters face: making actual quilt blocks. This problem requires us to apply our understanding of measurement ideas.

PART 1: Variations

Below are two variations of a star pattern.

1. Describe how the two patterns are alike and how they are different.
2. Make a copy of each pattern. Include the measurements of the sides of each figure.

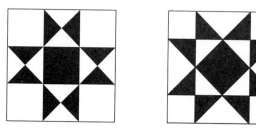

PART 2: Similar patterns

Examine the Cats and Mice pattern and the Eight Hands Round pattern.

1. How are they similar and how are they different?
2. Describe figures in the two patterns that are mathematically similar.

Cats and Mice   Eight Hands Round

PART 3: Challenges

Each of the following quilt blocks presents different challenges. For each block that you make:

1. Explain any quilt structures (such as patches, an 8-point star, or a hexagon).
2. Justify your choice of dimensions for the whole block (not all are squares).
3. Give step-by-step directions for making the figure. Include the length of each segment and degrees (when appropriate).
4. Describe your biggest challenge and how you overcame it.
5. Describe any insights, observations, what ifs, or questions that you have.

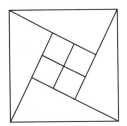

Arabic Lattice

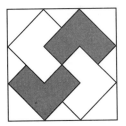

Card Tricks

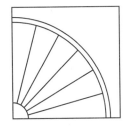

Grandmother's Fan[1]

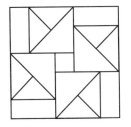

Hope of Hartford[1]

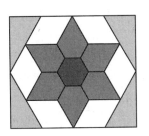

Diamonds in the Sky

Night and Day

Rambler

No. 364

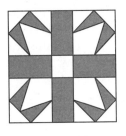

Crosses and Star

Irish Puzzle

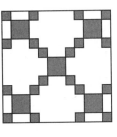

Puss in the Corner

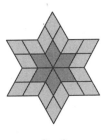

Star 2

PART 4: Quilting and systems of measurement

When quilting was invented, quilters used the English system of measurement. Because of this, many of the computations are cumbersome. Suppose you heard of a new quilting book with lots of great designs in which all the measurements were metric. Describe the pros and cons of switching to metric quilting.

[1] From Luanne Seymour Cohen, *Quilt Design Masters* (Palo Alto, CA: Dale Seymour Publications, 1996).

EXPLORATION 10.12   **How Much Will the Carpet Cost?**

I have seen problems like the following in many, many textbooks:

> Betty and Bruce have decided that the old carpeting in their living room has to go. They saw an advertisement on television for carpeting at $15.95 per square yard. Their living room measures 25 feet by 18 feet. How much will the new carpet cost?

The correct "mathematical" answer is obtained by the following process:

1. Find the area of the room: $25 \cdot 18 = 450$ square feet.
2. Convert from square feet to square yards: 450 square feet ÷ 9 square feet/square yard = 50 square yards.
3. Multiply this area by $15.95: (50 square yards)($15.95/square yard) = $797.50.

There is a major problem with this process: This is not at all how the carpet store determines how much your carpet will cost.

1. Brainstorm how the carpet store might determine the cost. Imagine that you are the manager of the store and/or that you are the person who installs the carpet. Summarize the results of your brainstorming as statements and/or questions.
2. After the class discussion, summarize the relevant information you will use to determine the cost of the carpet.
3. Now determine how much the carpet store would actually charge Betty and Bruce.

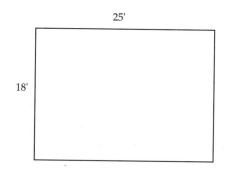

4. Determine the cost of carpet for each of the rooms below.

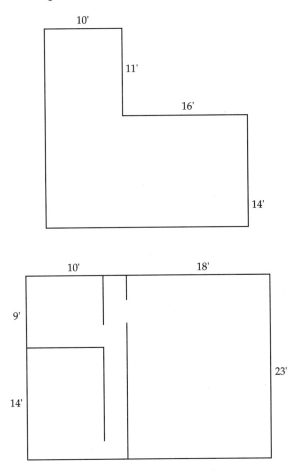

Looking Back on Exploration 10.12

Describe any questions you have about carpeting at this point.

EXPLORATION 10.13   **Irregular Areas**

Outside the classroom, when we want to find an area, more often than not the object is not one for which we have a nice, simple formula. However, we can use our knowledge of area to determine the areas of these objects. Applying your understanding of what area means, determine which of the two ponds below is bigger.

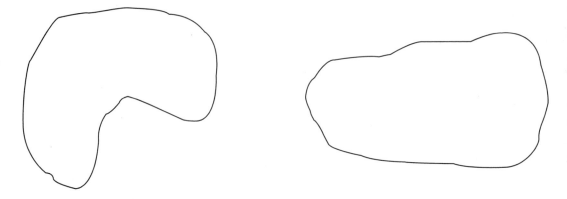

1.   Brainstorm different ways to determine which of these two ponds is bigger (feel free to use your imagination). Briefly summarize the methods that arose in your group.

2.   **a.**   During the class discussion, make notes of the pros and cons of each method, using a table like the one below. If you think any of the methods are not valid or will not be reasonably accurate, describe those concerns in your "cons."

	Pros	Cons
Method 1		
Method 2		
etc.		

      b.   Note any other important statements or points made in the class discussion.

3.   Select two methods that could be used to determine the actual area and determine the area of each of the ponds using that method. For each method:

      a.   Describe your work.
      b.   Justify the precision in your answer — that is, the choice of units, and the number of decimal places, if any.
      c.   Describe any difficulties you had and how you overcame them.
      d.   Describe any changes or modifications in your ideas with respect to the methods discussed in the previous step.

Looking Back on Exploration 10.13

Describe any questions you have about measuring the areas of irregularly shaped objects at this point.

EXPLORATION 10.14 **Functions, Geometric Figures, and Geoboards**

Throughout this book, we have seen that there are many functional relationships in mathematics and in everyday life. In this exploration, we will discover relationships among three variables: the area of a figure, the number of pegs on the border of the figure, and the number of pegs in the interior of the figure.

PART 1: Exploring the relationships between pegs and right triangles

1. Use your geoboards or dot paper (supplied at the back of the book) to fill in the top table on page 347. When a member of the group notices a pattern, discuss and note that pattern. Does that pattern lead to a hypothesis? For example, can you predict the area or number of border pegs or interior pegs in the next triangle?
2. When you feel that you are able to predict the area and the number of border and interior pegs when you are given only the dimensions of the triangle, state your hypothesis and how you came to discover it.
3. Predict the values for the three columns of the table for a 16 by 3 triangle and a 20 by 3 rectangle. Explain your predictions.
4. a. Use the top grid on page 349 to show the relationship between the length of the longer leg and the area of the triangle.
    b. What does this graph's being a straight line mean?
5. Use the bottom grid on page 349. In this step you are going to graph two different relationships on the same graph.
    a. First plot the points representing the relationship between the length of the longer leg and the number of border pegs. Then plot the points representing the relationship between the length of the longer leg and the number of interior pegs.
    b. What does neither of these graphs' being a straight line mean? Justify your response.
    c. Does it mean that they are not functions? Justify your response.
    d. Describe patterns that you see in these two graphs.
    e. Can you make use of these patterns to extend the graphs—that is, to predict the number of interior and border pegs when the longer base is 12, 13, 14, 15, or more units long?
    f. State your hypothesis now for predicting the number of interior and border pegs for any right triangle. Justify your hypothesis.

PART 2: Expanding our exploration to other figures

We are going to expand our exploration of relationships among border and interior pegs to include other figures. We are also going to change the focus slightly. Now, we are going to look for patterns so that we can predict the area of any geometric figure on the geoboard from the number of border and interior pegs.

1. Make several geometric figures on your geoboard or dot paper. In each case, determine the area (A), the number of border pegs (B), and the number of interior pegs (I) and record those values in the bottom table on page 347. Do you notice any patterns that lead to hypotheses? If you do, record your hypothesis and how it came to be. Test your hypothesis. If it works, great. If not, back to the drawing board. If you do not see patterns that lead to hypotheses, move on to another geometric figure.

2.   Now that you have determined the relationship among the three variables B, I, and A, look back and describe the moment of discovery.

Tables for EXPLORATION 10.14,   PART 1 and PART 2

PART 1: EXPLORING THE RELATIONSHIPS BETWEEN PEGS AND TRIANGLES

1.

Dimension of the right triangle	Area	Number of border pegs	Number of interior pegs	Insights/Observations
3 by 3	4.5 sq. units	9	1	
4 by 3				
5 by 3				
6 by 3				
7 by 3				
8 by 3				
9 by 3				
10 by 3				
11 by 3				

PART 2: EXPANDING OUR EXPLORATION TO OTHER FIGURES

1.

Geometric figure	Area (A)	Number of border pegs (B)	Number of interior pegs (I)

Grids for EXPLORATION 10.14,   **PART 1**

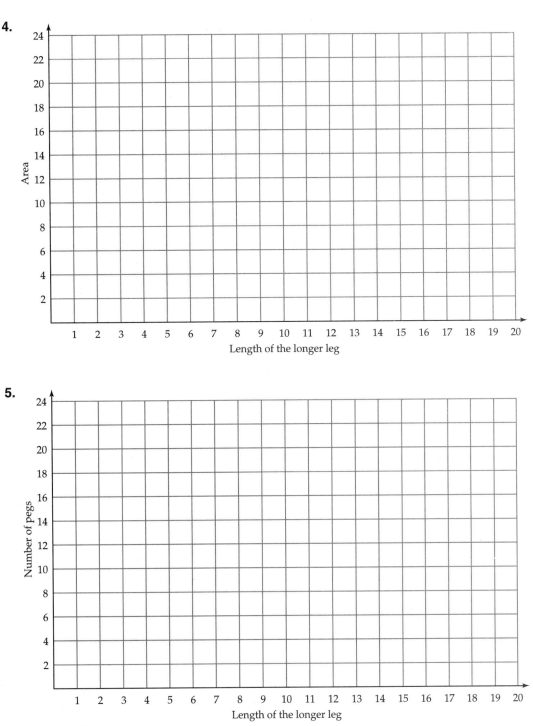

4.

Area

Length of the longer leg

5.

Number of pegs

Length of the longer leg

SECTION   **10.3**   **EXPLORING SURFACE AREA AND VOLUME**

Many applications of measurement involve three-dimensional surfaces—either covering them (surface area) or filling them (volume).

EXPLORATION 10.15   **Understanding Surface Area**

1. How would you determine the surface area of each of the objects below? Write down your thoughts.

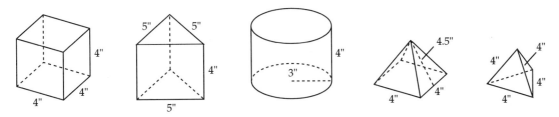

2. Meet with your partners and pool your ideas. Write down those ideas that the whole group understands and agrees with. The purpose of this step is not for someone who knows the formula to explain it to others so that they can copy it down. The purpose of this step is to discuss ideas and then note those ideas that make sense to the whole group.

3. Each person in the group will now construct a net [see Exploration 8.8 (Understanding Solids), Part 3] to make one of the models above. Remember that a net is a connected two-dimensional figure that can be folded up to make a space figure. Test each net by folding it to make the shape.

4. Each person presents a net and describes it to the rest of the group.

5. Now describe how to find the surface area of each of the figures.

6. **a.** Describe the commonalities and differences among the first three figures, using a table like the one below.

Commonalities	Differences

b. Describe a general formula for determining the surface area of any *prism*.

Looking Back on Exploration 10.15

Describe any questions you have about surface area.

EXPLORATION 10.16   **Understanding Volume**

PART 1: Volume of a cube

1. You probably know the formula for determining the volume of a rectangular prism: length · width · height. Can you justify this formula—that is, can you explain why we multiply these three numbers? Write your first-draft ideas.

2. Your instructor will give you a rectangular prism to fill with cubes. Do so and then revisit your response to the previous question. If you wish to revise it, do so now, and also explain what happened that led you to decide to revise your answer.

3. A manufacturer sells widgets. Each widget is put in a box whose dimensions are 4 cm × 6 cm × 10 cm. The shipping box is 80 cm × 60 cm × 40 cm. If the company makes 100,000 widgets each year, how many shipping boxes does it need?

    a. Using your knowledge of volume and your problem-solving tools, begin to work on this problem.
    b. Discuss your ideas with your partner(s). Remember to treat ideas as hypotheses: They may or may not be valid. Therefore, the purpose of the group conversation is to convince yourselves of the validity or nonvalidity of hypotheses that are raised.
    c. Describe your solution path and answer.

4. Let's say the company also manufactures gizmos. Suppose the dimensions of the box containing a gizmo are 3.5 cm × 7.6 cm × 11.3 cm, and your shipping box is 80 cm × 60 cm × 40 cm. How should the company arrange the gizmo boxes so that each shipping box will contain the greatest possible number of gizmos?

    a. Using your knowledge of volume and your problem-solving tools, begin to work on this problem.
    b. Discuss your ideas with your partner(s), remembering to treat ideas as hypotheses [as in Step 3(b)].
    c. Describe your solution path and answer.

5. Let's say each object is placed in a box whose dimensions are a, b, and c, and the dimensions of the shipping box are x, y, and z. Tell under what conditions the following formula will work for determining the number of small boxes that one shipping box will contain:

$$\text{Number of objects} = \frac{xyz}{abc}$$

    a. Using your knowledge of volume and your problem-solving tools, begin to work on this problem.
    b. Discuss your ideas with your partner(s), treating ideas as hypotheses.
    c. Describe your solution path and answer.

6. The United States Postal Service sells boxes. At the time of the writing of this book, these were the dimensions and prices of the boxes:

8″ × 8″ × 8″	$1.30
15″ × 12″ × 10″	$2.10
18″ × 12$\frac{1}{2}$″ × 3″	$1.90
20″ × 14″ × 10″	$2.50
Square Tube 3″ × 3″ × 24″	$2.05
Square Tube 3″ × 3″ × 36″	$2.50

    a. Are the prices of the boxes proportional to the volumes? Use your knowledge of volume and proportion and your problem-solving tools to work on this problem. State and justify your answer.

    b. Are the prices proportional to the surface areas? State and justify your answer.

    c. Which do you think is the more important measure in determining the cost of the boxes—surface area or volume? Justify your response.

PART 2: Volume of a cylinder

1. By now you should be comfortable with the formula for the area of a rectangular prism. What about the volume of a cylinder? Because its base is not a rectangle, we cannot determine its volume by filling it with little cubes. We must use reasoning to develop and then test our conjectures with respect to what its volume might be. Look back at the cylinder that you made from a net in Exploration 10.15. Describe your present thoughts concerning how to find the volume of that cylinder. Justify your reasoning.

2. There are several activities that you can do to justify and better understand how to find the volume of a cylinder. Below are several possibilities that will serve as hints. Explore one or more of these hints, and then summarize what you did and what you learned from the exploration.

    a. Look at the commonalities among the prisms. Imagine making prisms in which the base has more and more sides.

    b. Mark lines on your cylinder at 1 inch high, 2 inches high, and so on. Imagine slicing the cylinder at these points, so that your cylinder consists of four of these slices.

    c. Get a stick of modeling clay and make two congruent rectangular prisms that are 4 inches by 4 inches by 1 inch in height. Take one of those prisms and remold it so that you have a cylinder that is 1 inch tall. Obviously you have the same volumes. What does this suggest about how to find the volume of a cylinder?

3. State the formula for finding the volume of a cylinder and provide your justification of this formula in your own words.

PART 3: Volume of a pyramid

1. Look back at the models of a cube and a pyramid that you made from nets in Exploration 10.15. Describe your current thoughts concerning how to find the volume of that pyramid. It is important to note that "your current thoughts" is not the same as "your guess as to the formula." Rather, it is your thoughts about how we might find the volume of a pyramid. Your thoughts may contain some quantitative thinking—for example, recall Investigation 10.4, concerning the area of a circle, where we found that the area of a circle must be less than $4r^2$.

2. If you look closely at the pyramid and the cube, you will find some commonalities. If you haven't noticed them yet, examine these figures and discuss them now in the group. Consider how they might provide more clues about the connections between determining the volume of a cube and determining the volume of a pyramid. Write down your thoughts.

3. Your instructor will provide you with some material. Fill your pyramid and see how much greater the volume of the cube is than that of the pyramid. Before doing so, state your guess and any reasoning. Then do so.

4. Describe your present thinking about the relationship between the volume of a cube and the volume of a pyramid. State your thinking as carefully as possible. For example, it would not be accurate to say that the volume of a cube is greater than the volume of a pyramid; to be accurate, such a statement would have to indicate the conditions under which this would be true.

Looking Back on Exploration 10.16

Describe any questions you have about volume at this point.

EXPLORATION 10.17 **Applying Volume Concepts**

Let's say you started a company that manufactures a healthful, natural cereal. The dimensions of your cereal boxes are 24 cm by 16 cm by 5 cm. However, many customers have asked that you make a bigger box. Design a box that is similar in shape that will hold exactly 3 times as much cereal.

Your report of your findings needs to include:

1. Your actual work (not to be graded; may be messy).
2. Your answer.
3. Your degree of confidence about your answer and a brief explanation of your confidence; if your degree of confidence is not very high, state your uncertainty as questions.
4. A narrative description of how you arrived at your answer. This description may be a "tour" through your work.
5. A description of the biggest problem you encountered and how you overcame it.

EXPLORATION 10.18   **Determining Volumes of Irregularly Shaped Objects**

PART 1: Getting started and making predictions

1. **a.** When you go to the store, which size egg do you buy: medium, large, extra large, or jumbo? Why?

    b. Why don't we call the smallest size egg *small* instead of *medium*?

2. Are you getting the same value for your money with different-sized eggs? That is, would an egg that is twice as big cost twice as much? Write down your initial thoughts.

3. In mathematical terms, what would it mean to say that you are getting the same value when you buy a medium egg as when you buy a jumbo egg?

4. Do you think each of the sizes will be about the same value, or do you think they will vary (remember the 8-inch and 16-inch pizzas from Investigation 10.5)? Explain your reasoning. If you think the value will vary, predict the rank in terms of best value first, and explain your reasoning.

PART 2: Volume and cost

1. **a.** Predict the variation you would expect if we took 100 eggs of the same size and found their volumes. For example, what would be the ratio of the volumes of the largest and smallest eggs? Explain your prediction.

    b. If you were to graph the data with the volume as the independent variable and the frequency of each size as the dependent variable, do you think the graph would be uniform, normal, bimodal, or random? Explain your prediction.

2. Determine the volumes of the eggs and then determine the average volume for each size of egg.

3. Determine the relative value for each size. That is, are you getting approximately the same value for your money in each case?

PART 3: Mass and cost

1. **a.** Predict the variation you would expect if we took 100 eggs of the same size and found the mass of each. For example, what would be the ratio of the masses of the largest and smallest eggs? Explain your prediction.

    b. If you were to graph the data with the mass as the independent variable and the frequency of each size as the dependent variable, do you think the graph would be uniform, normal, bimodal, or random? Explain your prediction.

2. Determine the masses of the eggs and then determine the average mass for each size of egg.

3. Determine the relative value for each size. That is, are you getting approximately the same value for your money in each case?

PART 4: Mass and volume

1. Do you think the masses and volumes of different eggs will be proportional? First, explain in your own words what the question means. Then write down your hypothesis and explain your reasoning.

2. Using the data from Parts 2 and 3, determine the answer to the question from your data.

   Extension question: How do you think eggs are actually sized: by mass, volume, length, width, or something else?

Looking Back on Exploration 10.18

Describe any questions you have about this exploration.

EXPLORATION 10.19   **Paper Towels**

PART 1: Initial questions and ideas

Most households in the United States use paper towels. Putting aside for the moment the issues of conservation and waste, let's say you were in a supermarket and you were deciding which paper towel to buy.

1. What are some of the factors that might enter into your decision to buy a particular brand of paper towel?
2. How would you determine which brand of towel is the "cheapest"?

PART 2: Absorbency

1. Let's say you are more interested in absorbency than in anything else. That is, you want the towel that will absorb the most liquid. How might you detemine the most absorbent towel? Write down your initial ideas. After discussing ideas with your partner(s), develop a plan and then determine the most absorbent towel.
2. Your report must contain:

    a. Your plan and a justification of each step.
    b. Your actual data and conclusion. The reader must be able to see what you did and how you arrived at your conclusion.
    c. Your justification of the precision in your answer — for example, the choice of unit and the number of decimal places, if any.
    d. A description and explanation of your degree of confidence in your result.
    e. A description of any difficulties you had and how you overcame them, either totally or partially.

3. Present your findings to the class. After the class presentations, describe and explain any changes you would make in your design.

PART 3: Strength

1. Let's say that you are more interested in strength: You want a towel that won't shred when you clean surfaces with it. How might you determine the strongest paper towel? Write down your initial ideas. After discussing ideas with your partners, develop a plan and then determine the strongest towel.
2. Your report must contain:

    a. Your plan and a justification of each step.
    b. Your actual data and conclusion. The reader must be able to see what you did and how you arrived at your conclusion.
    c. Your justification of the precision in your answer — for example, the choice of unit and the number of decimal places, if any.
    d. A description and explanation of your degree of confidence in your result.
    e. A description of any difficulties you had and how you overcame them, either totally or partially.

3. Present your findings to the class. After the class presentations, describe and explain any changes you would make in your design.

EXPLORATION 10.20　**Measurement, Ambiguity, and Precision**

Most textbook problems are well defined, but many, if not most, real-life problems are not well defined. Below are six problems that are not well defined. In some cases, you need to gather more information in order to solve the problem. In all cases, you need to make (and justify) assumptions in order to solve the problem.

Each group will select a question, determine a solution, and present the group solution to the class.

Your report of your findings needs to include

a. Your solution in a two-column format:

- The left column contains the actual mathematical computations.

- The right column describes the assumptions you made in order to solve the problem and justifies/explains why you are doing what you are doing.

b. Your degree of confidence about your answer and a brief explanation of your confidence; if your degree of confidence is not very high, state your uncertainty as questions.

c. A description and justification of the assumptions you made in order to solve this problem.

d. A description of the biggest mathematical problem you encountered along the way and how you overcame it.

e. A description of your one or two most important learnings about mathematics and/or about problem-solving.

Questions:

1. McDonalds has now sold over 100 billion hamburgers. Tell us about how much 100 billion hamburgers is. At least one aspect should be to describe how much volume this many hamburgers would take up; you may, but are not required to, represent this answer as a cube.

2. It is estimated that approximately 16 billion disposable diapers are sold each year in the United States. How much volume does this amount of diapers represent?

3. Myra believes that it would be cheaper to make soda cans in the shape of prisms (like children's juice boxes) rather than the present cylinders. Do you agree or disagree with this premise? Describe the pros and cons, both mathematical and other, of making soda cans in the shape of prisms.

4. I once heard on the radio that Congress had just appropriated money to increase office space by 65,000 square feet and to add 18,000 square feet of parking. How many parking spaces can be gotten from 18,000 square feet? Draw a rough blueprint for the parking lot—total dimensions, length and width of parking spaces, the width of the "lanes," and so on.

5. One of the things I miss most about high school teaching is that in high school I had my own room and could decorate it as I wished. College teachers do not have their own classrooms, and so the walls of most college classrooms are rather bare. One day I thought that if we could cover the walls of each college classroom with corkboard, it would make posting things on the wall easier. I wondered how much this would cost. How much would it cost to cover the walls of your present classroom with corkboard?

6.   If we took all the office and writing paper that is thrown away each year, we could build a wall from New York City to Los Angeles that was $8\frac{1}{2}$ inches wide and 7 feet high. If we took all that paper and dumped it onto your campus, how high would that pile be?

"Waste wall: Coast-to-coast," from *USA Today*. Copyright © 1991, *USA Today*. Reprinted with permission.

INDEX

Accuracy, 327
Acute triangle, 241
Adding up algorithm, 59–60
Addition
  in Alphabitia, 55–57
  alternative algorithms, 59–61
  of decimals, 165, 166
  of integers, 130
  mental, 93, 95
  models of, 152, 153
  properties of, 313
Addition table, in Alphabitia, 57
Additive comparisons, 171, 185
Additive identity, 313
Additive inverse, 131, 313
Algebraic thinking, 23
  border growth pattern, 31–33
  figurate numbers, 27–30
  magic squares, 8–9
  pattern block patio, 35–36
  relationships between
    variables, 24–26
Algorithms, alternative
  addition, 59–61
  division of fractions, 154
  division of whole numbers,
    84–86
  multiplication, 76–79
  subtraction, 62–63
Algorithms, standard
  division, 82–83
  multiplication, 75
Alphabitia
  addition in, 55–57
  fractions and, 136
  numeration system, 37–41, 43
  story problems in, 56, 58
  subtraction in, 58
Angles, of tangram, 330
Answers, justifying, 13
Area
  of circle, 339
  cost of carpet and, 342–343
  on geoboards, 332–337
  irregular, 344
  meaning of, 331
  of parallelogram, 334
  perimeter and, 336–337
  of square, 332
  surface area, 351–352, 360
  of tangram pieces, 330
  of trapezoid, 335
  of triangle, 334–335
Area models
  of fractions, 141–145
  of multiplication, 152, 153

Arrows, tessellations of, 295
Associative property, 313, 314,
  316
Assumptions, 360–361
Average, 190–191

Bar graph, 188
Base, of numeration system,
  43–49, 72
Base 2, 44, 46–48
Base 3, 45
Base 5, 43, 46–48
Base 6, 43–44, 46–48
Base 10, 43
  circle clocks, 71–72
  translating from, 47–48
  translating to, 46–47
Base 10 blocks
  for decimals, 156
  for division, 82
Base 12, 44–45, 46–48
Borrowing, 58, 62, 63
Boxes, volumes of, 353–354, 356
Boxplot, 200
Brahmagupta, 154, 155
Bringing down, 83, 84
Buffon's needle, 198–199

Carrying, 58, 63
Cartesian product model, 152
Census, 184–185
Center, measures of, 190–191,
  192, 194
Chance, see Probability
Checking answers, 13
Children's thinking, about
  subsets, 15–16
Circle
  area of, 339
  radius of, 339
Circle clocks, 71–72
Circle graph, 188
Combinatorics, 203–207
Combining, addition as, 152, 153
Communication
  about geoboard figures, 210
  about new shapes, 237–240
  about three-dimensional
    figures, 243, 244, 245, 246
Comparison model, of
  subtraction, 132, 152
Comparisons
  additive, 171, 185
  multiplicative, 171, 185
  of prices, 4, 171
Composite numbers, 111

Concrete operational thinking,
  15
Cones, cross sections of, 250
Congruence
  with geoboards, 215
  with tangrams, 224
Congruent vertex point, 295
Convention, 119
Cooperative learning, 3
Cross product algorithm, 77
Cross sections, 248–251
Cubes
  cross sections, 248, 249
  nets for, 233–234, 246–247
  from polyominoes, 233
  surface areas, 351–352
  two-dimensional
    representations, 242–244
  volumes, 353–354
Cuisenaire rods, fractions and,
  141
Cylinders
  cross sections, 249–250
  nets for, 247, 351
  surface areas, 351–352
  volumes, 354

Darts, 5
Data
  analyzing with sets, 18–19
  collecting, 18, 24–25, 190, 191,
    192, 194
  distributions of, 190–194, 198,
    200–201, 357
  graphing, 183–189, 190–191,
    192, 200, 357
  measures of center, 190–191,
    192, 194
  reliability of, 183, 189
  representations of, 19, 21
  survey, 194
  validity of, 183, 189
Decimals, 156–168
  with base 10 blocks, 156
  estimation with, 158–163
  operations with, 165, 166–168
  place value in, 157
  precision with, 359
  target games with, 157,
    166–168
  see also Fractions
Decomposition
  of fractions, 135
  of whole numbers, 105, 118
Demiregular tessellations, 295
Denominator, 136, 138, 143

CUTOUTS

Base 10 Graph Paper

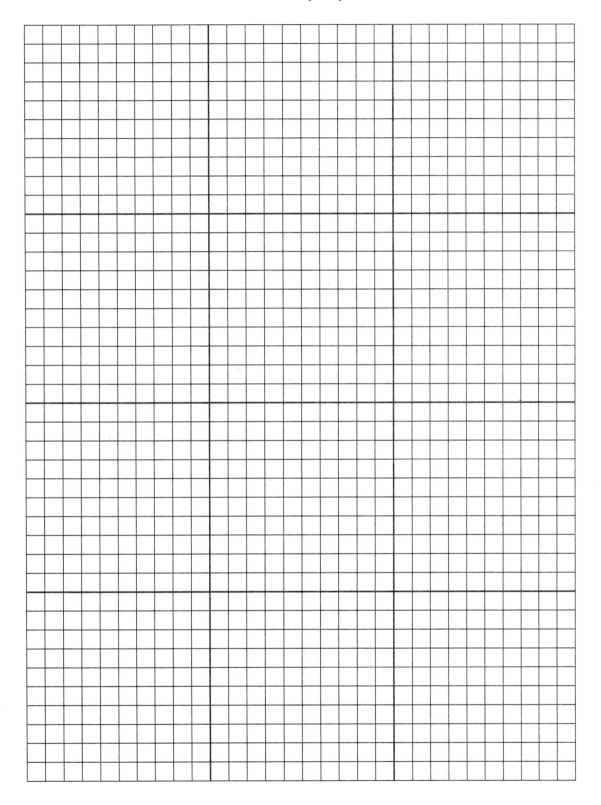

Other Base Graph Paper

Other Base Graph Paper

Other Base Graph Paper

Other Base Graph Paper

Geoboard Dot Paper

Isometric Dot Paper

Polyomino Grid Paper

Polyomino Grid Paper

Polyomino Grid Paper

Polyomino Grid Paper

Tangram Template

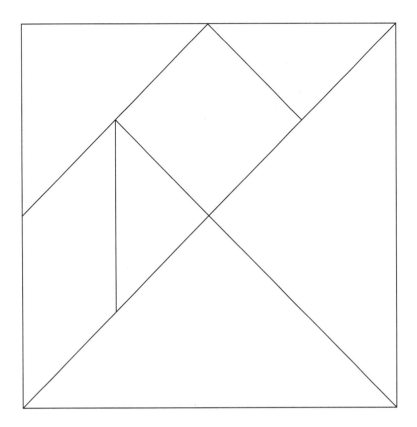

Regular Polygons

Exploring the Area of a Circle

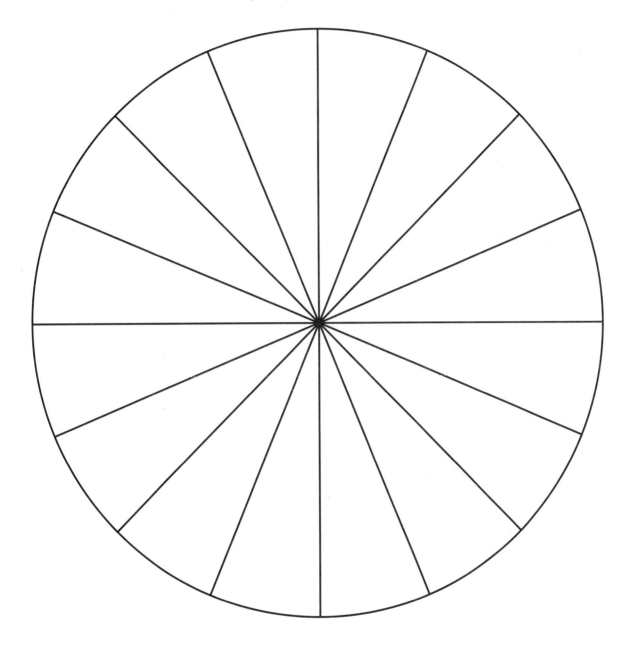